SERVIC
A HANDBO

Other books by John A. Murphy:

Hygiene in Practice

Quality in Practice

Service Quality in Practice

A Handbook for Practitioners

John A. Murphy
with Tony Farmar

GILL AND MACMILLAN

Published in Ireland by
Gill and Macmillan Ltd
Goldenbridge
Dublin 8
with associated companies throughout the world

0 7171 1879 7
Index compiled by Helen Litton
Print origination by
Seton Music Graphics Ltd, Bantry, Co. Cork
Printed by Colour Books Ltd, Dublin

A catalogue record is available for this book from the British Library.

To my wife Eleanor
and children Niall, Maria and John Gerard

Diagrams and Case Studies

Diagrams and case studies, largely from the experience of Irish companies, have been inserted throughout the text. For practical reasons they do not necessarily relate to the content of the chapter in which they occur.

Contents

Acknowledgments

Few books today contain solely the input of one person. In writing this book I am especially grateful to my family for their unselfish support, understanding and patience. The overall support of Telecom Éireann and in particular its executives notably Jim Lynch and Michael G. Ryan is acknowledged with sincere gratitude. I am also indebted to the following organizations and individuals for their case-study contributions.

Aer Lingus — Colm Moran; Alfred K. O'Hare & Company Ltd — Joseph O'Hare; American Express Ireland Ltd — Shelagh Twomey; Bank of Ireland — Cormac O'Herlihy; The Educational Company of Ireland — John V. Bowers; Electricity Supply Board — Ken O'Hara; IBM Ireland Ltd — Bernie Trench; Northern Telecom (Ireland) Ltd — Dee Goggin; Patterson Kempster & Shortall — Norman V. Craig; Superquinn — Derrie O'Reilly; Texaco — Ray Quilligan; Ulster Bank Ltd — Noel Kincade; VHI — Paddy Creedon; and Patrick Casey — optometrist.

Among those who gave generously of their time reviewing and commenting on sections of the initial typescript were Michael Wallace of the University of Limerick, Paddy White of Intel Ireland and Dan Boland of Boland Quality Management, whose reading of the entire manuscript was reassuring. I also owe special thanks to Phyllis Kavanagh for her administrative assistance and my other colleagues in the Irish Quality Association, Eleanor Brodie, Barbara Hughes, Julie Powney, Jacqueline Cryan and Jack Carroll, whose overall commitment, dedication and professionalism made it possible for me to embark on this task. Also to my colleagues at the University of Ulster, Professor Nial Cairns and Professor Richard Harrison for their advice and assistance.

Particular appreciation must be given to Tony Farmar who has been involved in all aspects of planning, developing and editing of the book. Thanks are also due to Gill and Macmillan Ltd for their assistance and continuing confidence in my work.

Finally I extend my deepest gratitude to my wife Eleanor for her ongoing support. Without her constant interest, support, patience and encouragement over the years, *Service Quality in Practice* would not be a reality.

CHAPTER 1

Introduction

'You don't have to do this:
survival isn't compulsory.'
W. Edwards Deming

Service Quality in Practice is intended as a practical handbook to the total approach. It is for those who have decided on the total service quality approach already, and want a guidebook, rather than those who are looking to be inspired or converted.

Six key areas of strategic focus are identified. These are:

- client focus.
- employee involvement.
- understanding the process.
- measurement.
- quality assurance systems.
- continual improvement.

These six elements are the essential focus points of any quality policy. They are not in themselves a plan; however, they do identify areas which any satisfactory quality plan must address.

Throughout the book, case studies of quality management practice in real Irish service organizations, supplied by themselves, illustrate the points being made. These case studies take up specific problems or aspects of service quality management and present the solution arrived at by an Irish-based company. The book ends with a chapter on how to implement total service quality.

This book is written to help companies to develop service quality management strategies that will enable them to meet the needs of the 1990s. The industrial experience since the Second World War bears out the importance of this objective.

Firstly there was the success of the Japanese. They seemed to be able to target well-established business areas, and then by consistently delivering quality products, to dominate market after market. Word seeped from Japan that the secret of this success was in the quality circles and other quality techniques. This led to the adoption of these techniques as a quick fix in fashionably minded companies. What these companies had not realized was the need for corporate culture change at the same time: outside Japan the quality circle movement usually ran into the sand. People began to wonder if this kind of high level quality management was only possible in the unique circumstances of Japanese culture.

Then the American PIMS database confirmed the message. Quality was now to be the key to long-term success for American companies too. The PIMS database was started by General Electric in the 1970s as a method of discovering what factors led to greater profitability in companies. The preconceptions of its founders can be read in the name, which stands for Profit Impact of Marketing Strategy. The PIMS database, which now collects information on three hundred variables from over three thousand companies, is run by the Strategic Planning Institute of Cambridge, Massachusetts. Ironically one of the key findings of the research has been to downgrade marketing expenditure as a determinant of long-term return on capital.

This database is clearly a valuable information resource, of sufficient breadth to take in a wide variety of possible factors. There is however a question as to how widely applicable its findings are. The companies providing information to the database, and receiving detailed feedback, are paying substantial fees for this service. They are therefore a self-selected group of the most highly motivated companies, prepared to invest heavily in this sophisticated self-analysis. Such companies would take for granted the presence of, for instance, highly efficient corporate planning and cost control systems. The contributors to the database are therefore equivalent to Olympic sprinters; advice suitable for them may not be appropriate for more plodding runners.

Early studies on the PIMS database concentrated on the link between market share and profitability. There is still a strong relationship between the two. The higher the market share, the higher the return on investment. The studies then began to explore how market share might be developed. This sequence of thought is explained in *The PIMS Principles* by Buzzell and Gale:

'In the long run, the most important single factor affecting a business unit's performance is the quality of its products and services relative to those of competitors. A quality edge boosts performance in two ways:

- in the short run, superior quality yields increased profits via premium prices. As Frank Perdue, the well-known chicken grower, put it: "Customers will go out of their way to buy a superior product, and you can charge them a toll for the trip." Consistent with Perdue's theory, PIMS businesses that ranked in the top third on relative quality sold their products or services, on average, at prices 5–10 per cent higher (relative to competition) than those in the bottom third.

- in the longer term, superior and/or improving relative quality is the more effective way for a business to grow. Quality leads to both market expansion and gains in market share. The resulting growth in volume means that a superior-quality competitor gains scale advantages over rivals. As a result, even when there are short-run costs connected with improving quality, over a period of time these costs are usually offset by scale economies. Evidence of this is the fact that, on average, businesses with superior-quality products have costs about equal to those of their leading competitors.' (Buzzell and Gale 1987, 7)

As international markets become more sophisticated, producers and sellers of services and products will have to make equivalent improvements to their whole way of operation. A recent survey of international business people called *Service — the New Competitive Edge* carried out by John Humble for Management Centre Europe, revealed that 80 per cent of the 3,375 respondents consider that service quality is the new key to competitive success in the 1990s; 90 per cent of respondents expect service quality to become more or much more important in the next five years. Incidentally, 88 per cent of Japanese businessmen believed that service quality will become 'much more important' in the next five years. In this survey service was defined in the widest possible sense 'to reflect a company's service philosophy internally and externally, its service policies and procedures as well as all the individual elements that go to make up an overall service concept'.

A similar survey organized in Ireland in 1991 revealed that 81 per cent of Irish managers now see quality and service as the key to competitive success. A majority of managers (77 per cent of those questioned) recognized service quality as vitally important for business success now and in the future. About the same proportion also saw customer expectations as significantly increasing in the next few years. Constantly rising standards is one of the three key pressures driving service quality.

The other two are fierce competition from service-oriented companies, and the need to establish good service as a differentiator. Despite this theoretical understanding, in practice less than half of Irish firms prepare reports on customer satisfaction; just over half claim to have quantified performance standards. Training performance also lags behind theoretical allegiance. Only a quarter of the companies have provided training for those who actually serve customers as opposed to 44 per cent of companies' training managers. Both figures are low in a fast-evolving environment.

In this survey, which is based on results from 424 companies, managers identified five factors critically influencing purchasing decisions. These were (in order of importance):

- quality ('fitness for use as perceived by the customer').
- reliability ('meeting the customers' needs as promised').
- speed of delivery ('being responsive to customers' needs').

- courtesy ('respect for the customer as an individual').
- price.

The report comments: 'The Japanese have different ranking of these buying factors. They view reliability, quality and problem solving as the biggest factors influencing customers to buy. The rest of the world — including Ireland — rank quality, reliability, and problem solving in that order. Perhaps the Japanese view reliability as the future battleground, having already mastered quality.' (Quinn and Humble 1991)

This growing understanding of the importance of service quality reflects (a little late) the growth of service in the ordinary industrial economy. Sixty years ago service expenditures such as marketing and administration were small relative to manufacturing, which as a result got major attention. Today service elements have become increasingly important. Economists and tax gatherers distinguish service suppliers, such as lawyers, dentists, department stores and airlines, and manufacturing companies, i.e. those that produce transportable goods. In real life these distinctions appear arbitrary. Every organization is, at least partially, in the service business.

Any operation has varying degrees of both production content and service content. In some cases the product content is relatively high, in others quite low. However, whether you provide meals on wheels, run a hairdresser's salon, or manufacture heavy metal goods, the service element to the customers is critical to your success, and becoming daily more so. By the end of the century up to 90 per cent of employees in the US manufacturing sector will work in administrative and service functions.

This is not to say that there are not substantial differences between a service and a transportable good or commodity. Some of these differences are important in identifying the unique nature of service. Karl Albrecht, the American service quality guru, identified the following ten distinctive characteristics of service in his bestseller *Service America!* (Albrecht and Zemke 1985, 36–7). Not all of these apply to every service function, but cumulatively they paint a picture of the special transaction between buyer and seller implicit in the concept of service.

1. A service is produced at the moment of delivery; it cannot be created in advance or held in readiness. You cannot keep a stock of haircuts or legal opinions.
2. A service cannot be centrally produced, inspected, stockpiled or warehoused. It is usually delivered wherever the customer is, by people who are beyond the immediate control of management.
3. The 'product' cannot be demonstrated, nor can a sample be sent for customer approval in advance of the service.
4. The person receiving the service often receives nothing tangible; the value of the service depends on his or her personal experience.
5. The service cannot be passed on to a third party.

6. If improperly performed, a service cannot be recalled. Once the barrister has presented the case in court, and the judgment is declared, no replay is possible.
7. There is no opportunity for quality inspectors or other controls to intervene between a human error and the creation of an unsatisfactory service.
8. Delivery of the service usually requires human contact.
9. The receivers' expectations control their satisfaction. As a result objective evaluation of the service is very difficult.
10. The more people the customer must encounter during the delivery of the service, the less likely it is that he or she will be satisfied.

All this adds up to the creation in all companies of a Right First Time culture (see for instance *Creating Culture Change* by P.E. Atkinson, 1990). This does not of course apply only to product quality. Attention is increasingly being drawn to service quality. The pace of technical development has contributed to this. As products become more complicated, more attention must be paid to enabling the consumer to make effective use of them. There is little point in adding elaborate product enhancements if the consumer cannot understand how to operate them. This is happening more and more. A quarter of purchasers of video recorders do not know how to use them for recording programmes; most washing machine users rarely change the 'programme' on their washing machine; as much as 80 per cent of the capacity of most computer programs is unused. In each of these cases a sophisticated technical product is simply not providing full value to the customer because not enough effort has gone into delivering the product to the customer in a usable form. The product quality may be high, but the customer will never discover this because of the poor quality of service.

Everyday experience reinforces the point. Typically much less thought has gone into the service delivery aspects of all kinds of products than into the manufacture of the product itself. In America the poor quality of personal service has become commonplace; in Ireland too, despite our reputation for friendliness and our dependence on the tourist industry, little thought has gone into developing service quality standards. Such experiences as these (all from recent personal experience) are commonplace:

- a £750 quality control instrument is finally delivered two months after its promised delivery date.
- a laundry returns a coat after dry cleaning without its belt. The customer had to go back to retrieve the belt, and was shown a box of belts to choose from.
- a licensed taxi-driver smokes all the way from the airport. During the journey he engages in a shouting match with another driver.
- a specialist magazine on order is not available on the usual day; on returning two days later the customer is told that all the newsagent's stock was stolen the day before, and he will have to look elsewhere.
- the only convenient bank ATM is closed for the weekend.

In each of these cases the problem is not so much the technical system as the service. There is probably nothing wrong, for instance, with the ATM; it is just that someone has failed to feed it with enough notes. The taxi-driver gets his passenger to the required destination, but with as little grace as possible. Time and again customers experience proficient technical product quality combined with poor service. Complaints about professionals, for instance, are much more likely to be about their inefficiency than their incompetence.

However, although such events are commonplace, very often the management of the service organization does not hear of them because few customers with such experiences bother to complain. After long experience of poor service quality, most consumers have come to believe there is no point in complaining. This applies even in company-to-company transactions: the US Research group TARP found that as many as one-fifth of industrial customers failed to complain about a problem: of those, 70 per cent did not complain because they felt it wasn't worth the trouble, 20 per cent because the rep wasn't available and 10 per cent because they felt it would do no good. If the service was that poor to begin with, only a fool would suppose that it would get better without a fight. Life is too short to spend it improving the manners of service clerks, or explaining to doctors that the practice of block booking patients for appointments is not only inefficient for the patients, but also requires the practice to tie up space for waiting rooms that would be more usefully used for medical purposes.

The dissatisfied customer does however use that most powerful weapon, word of mouth, to spread the bad news. The organization might not get to hear of a problem which has arisen, but research suggests that ten people on average get to hear of small scale problems (involving a loss of £5 or less), and as many as sixteen of large problems. Naturally the people who are told are those likely to be interested, i.e. existing or potential customers.

Research in the US discovered that:

- 'the average business never hears from 96 per cent of its unhappy customers. For every complaint received, the average company in fact has 26 customers with problems, 6 of which are serious problems.
- the average customer who has had a problem with an organization tells nine or ten people about it. Thirteen per cent of people who have a problem with an organization recount the incident to more than 20 people.
- customers who have complained to an organization and have had their complaints satisfactorily resolved tell an average of five people about the treatment they received.
- complainers are more likely than non-complainers to do business again with the company that upset them, even if the problem is not satisfactorily resolved.
- of the customers who register a complaint, between 54 and 70 per cent will do business again with an organization if their complaint is resolved. That figure goes up to a staggering 95 per cent if the customer feels that the complaint was resolved quickly.'(Albrecht & Zemke 1985, 6)

Paradoxically, complainers are more likely than non-complainers to do business again with the company. Even if complaints were not resolved, complainers are more loyal than non-complainers. TARP, the source of these figures, found that for losses of over $100, 19 per cent of unresolved complainers declared loyalty while only 9 per cent of non-complainers did so. By contrast 82 per cent of quickly resolved complainers stayed loyal. The same relationship between the loyalty of satisfied and unsatisfied complainers holds in industry after industry. In financial service for instance, TARP found 73 per cent loyalty for satisfied complainers and 17 per cent for unsatisfied; in telecommunications the ratio was 93:38.

Complainers want to value their relationship with the company. In the old days the grander kind of customer used to talk about 'my solicitor', or 'my butcher, baker, candle-stick maker . . .', not that anyone was supposed to imagine that the speaker actually owned these people, but that there was a recognized and valuable relationship on both sides. When madam went shopping to the local grocer, a chair was provided while the items on her list were picked, discussed and packed, perhaps a biscuit was laid out for the dog, and finally the goods were delivered. The modern supermarket has to work very hard to match this level of service. In pursuit of 'efficiency', companies have tended to ignore this potentially valuable asset.

The best companies have recently recognized this, and have turned their attention to nurturing the relationship with the customer. However, the Management Centre Europe report cited above found that Irish companies are seriously deficient (compared with international competition) in six key areas. These are:

- understanding customer needs: we are not exploiting all possible means to explore customers' needs. There is also a strong dissatisfaction with those methods used. For instance 56 per cent used questionnaires, but only 19 per cent thought them effective; 71 per cent (particularly of firms employing 500 people or more) used market research, but less than half found it effective. This effectiveness proportion contrasts strongly with the Japanese experience. This suggests that the techniques used may not be adequately shaped or tailored to the job.

- measuring customer satisfaction: 85 per cent use customer complaints as a measure, though only two-thirds of those regard them as effective. Questionnaires, independent assessments and customer focus groups are used by about 40 per cent of Irish respondents, but again the number rating them as effective is low. One is entitled to suspect that if managers regarded customer satisfaction as highly as they claim, they would insist on more satisfactory methods of measuring it.

- setting performance standards: only 57 per cent of Irish managers claim to have set quantified performance standards on key service issues. Although this figure is low, it is considerably higher than the Japanese, only 36 per cent of whose companies have such standards.

- training all employees in customer service: we will see later in this book that the modern service oriented company does not treat employees as operatives, but as players actually on the pitch. Players score goals, not trainers. In this view, there is no sense in differentiating between management and non-management training.
- product quality is not enough: product quality attracts the customers, reliability keeps them. Quality will be the battlefield of the 1990s in Europe; if the Japanese are right (and they have been so far) the next attack will be on reliability. Irish firms must start to train for that now.
- exploiting information technology: information technology is the key weapon in improving customer service. Service environments are changing rapidly from low-tech to high-tech: in 1988 Tom Peters said that some service industries are now more technology based than giants such as US Steel. Ninety per cent of new computers go into service environments. This means developing good information systems, good electronic connections with customers, setting up sensing systems to initiate fast reactions, and enabling the players on the spot to make decisions.

Quality management began in the factory. In the early days of the century companies competed on product design and availability. Productivity was therefore the key to success. The so-called 'American system of manufacture', whereby products were designed with interchangeable parts, each of which was made by separate machines, stimulated this, replacing the ancient craft style whereby each worker made a whole product. Individual workers were now designated to manufacture only one part, perhaps never seeing where their part fitted into the whole at the end of the process. As a result motivation and skill were neglected, and so product quality suffered. Waste accumulated; in some factories as much as a quarter of the workers spent their days repairing work botched by the remainder.

In the face of this it was obviously prudent to develop some kind of control system. Techniques such as statistical process control, inspection sampling etc. were developed and became widespread. This was all perceived as a matter of the internal economy of the factory: it had little or nothing to do with the customer. The Japanese widened the ambit of the techniques to cover the whole company with their Total Quality Control concepts, and with Quality Function Deployment (also called Hoshin Planning) began to develop products that met customers' needs. The concept of quality management was still largely reserved for manufacturing organizations. If it was proposed that service companies should use the same disciplines, the managers would shake their heads tolerantly, and allude to the mysterious and unmeasurable qualities of courtesy, pleasantness and so on.

Quality of service can, however, be made amenable to quality assurance ideas. Not only *can* service organizations apply these disciplines, but it has become increasingly clear that they *must.* Consumers have had to put up with poor service quality from all types of organizations for too long. They have learned to tolerate it, if not to like it, on the lines of the old Arab proverb: 'pray God I will not have to put up with what I could endure'.

This is about to change. As service industry takes up an increasing proportion of the economy, and as consumers become aware that it *is* possible to get better presentation and service, the pressure to compete in this area will be irresistible. According to the statisticians, some 55 per cent of the Irish workforce are now in the services sector, 29 per cent in the manufacturing sector and the remainder (16 per cent) in the agricultural sector. This compares with 69 per cent in services in the USA, 66 per cent in the UK, and 57 per cent in Japan. Since 1961, employment in the services sector (including health but excluding defence and public administration) has gone from 365,000 to 550,000. (OECD Labour Force statistics, quoted in *Administration Yearbook and Diary*, 1989.)

It has been argued recently that this apparently enormous relative increase in service activity is not a sign of the sickness of manufacturing and the sturdy growth of demand for services, but the other way round. What is actually happening is that manufacturing activity is becoming increasingly efficient, and so requires less resources for the same output. Service activity on the other hand has failed to create similar efficiencies. Until very recently, there had been very little change in basic styles and technologies of service operations for thirty years or more. Few manufacturing operations could say the same. As a result, as J.W. O'Hagan's standard text *The Economy of Ireland* points out, service employment in Ireland between 1970 and 1987, went from 43 per cent to 57 per cent of the workforce: at the same time the contribution to GDP fell from 55 per cent to 53 per cent (O'Hagan 1991). If this argument is correct, it is yet another reinforcement of the need to introduce quality management techniques into the services sector.

In the international standard guideline ISO 9004 Part 2, the *Guide to Quality Management in Service Industries*, the approach adopted to the management of service quality is straightforward. The service process is divided into a series of discrete stages. In a fully developed system, each of these stages should be documented by a quality manual describing the general process and standard operating procedures giving the more changeable basic detail. In the international standard, the service process is broken down into four stages:

1. Producing a service brief: this process combines market research, awareness of customer needs, feedback from customers, internal assessments, legal and commercial obligations, and other inputs to produce the Service Brief, which 'defines the customer's needs and the related service organization's capabilities as a set of requirements and instructions that form the basis for the design of a service'.
2. Designing the service: once the service brief (which as we have seen defines what the customer wants) is in place, the next stage is to convert the brief into design specifications for the service itself and how it is to be delivered.

This includes assigning responsibility for each stage in the service process and developing quality control procedures based on standards of acceptability for each service characteristic. The specific stages of the process such as receiving information from clients, purchasing from subcontractors, delivering the service itself and finally invoicing are all included in this stage, which should be done with the aid of flow charts.

3. Delivering the service: the delivery of the service entails adherence to the specification and systems outlined in stage 2 above, combined with monitoring that the specification is being met and coping with any deviations. This requires continual internal and customer monitoring of the delivery process against specification, and the establishment of an effective corrective action system to ensure that non-conformances are designed out of the system.
4. Service performance analysis and improvement: the corrective action system may throw up problems that cannot be solved immediately. They may require more fundamental revisions. A key part of any quality system is the continuous improvement strategy. There should be an unceasing effort to identify the key service characteristics and how they might be improved. In today's marketplace, any product/service combination that is the same as it was last year is almost certainly not meeting present needs.

The standard guideline ISO 9004-2 is designed to be used in developing and analyzing the quality management systems of organizations largely or mainly involved in service, such as hotels, hospitals, insurance companies, professional practices etc. However, its potential range of use is much wider than that. As we have seen, all manufacturing companies necessarily have a service element in their product, but there is a more fundamental and widespread application of this service quality approach.

In his latest book *Building a Chain of Customers*, Richard Schonberger, the proponent of World Class Manufacturing ideas, argues that business functions should be linked to form a continuous chain of customers, from one function to the next. In this formulation, everyone both is and has a customer; whether of the next department, the next office or the next factory (Schonberger 1990). Exactly the same controls and disciplines should operate between any supplier and its customer, whether internal or external.

The objective of this book is to enable companies to deliver a superior relative product quality that will lead to an increased market share and a successful long-term return on capital. The phrase we will use to describe this quest for quality, an analogy with manufacturing's Total Quality Management and other similar phrases, is Total Service Quality. Each of the words in the phrase has a special significance.

Quality is defined, as it usually is in this context, as the meeting of customers' expectations, as conformance to requirements. It does not mean providing Café de Paris service in the local cheap and cheerful chipper. Customers' expectations are the key to quality: in the Café de Paris customers expect (and are prepared to pay for) a standard of attention, an imaginative and stylish menu, a superfluity of linen and cutlery that would be absurd in the chipper. Yet it is as easy to get pretentious, precooked food in the Café de Paris as it is to get delicious rasher and eggs in the chipper. Quality is not a measure of pretensions, it is a measure of meeting expectations.

The other two words in the phrase 'Total Service Quality' are equally important. Service describes exactly how these expectations are delivered. This is as valid in heavy manufacturing as in hairdressing.

The word 'Total' signifies that the whole company is involved. Delivering the customers' expectations is not just a matter for the salespeople or the front-line operations. Every member of the staff must understand how their part of the operation delivers some value to the customer. This is not a trivial point. The analogy might be drawn with warfare. Before Napoleon, wars were fought by professional soldiers who were often mercenaries: the ordinary people were little involved. Jane Austen, writing in the middle of the Napoleonic wars, makes no mention of them either in her novels or her letters, despite her brothers' involvement in the fighting Navy. Since the American Civil War, however, wars have involved every citizen on both sides: wars have become total. The same degree of total involvement is required to meet the expectations of the modern consumer.

ACTION

- **list all the incidences of poor service you have suffered recently.**
- **did you complain?**
- **did you tell anyone else?**
- **how many of Albrecht's ten elements apply to your service?**

SUMMARY

Every organization provides both a service and a product. As products and consumers become more sophisticated, they require a large input from service elements to make them effective. Many business writers now believe that quality of service is the key business battleground of the 1990s. Service Quality in Practice *is designed as a practical handbook; it will help your company to create the Total Quality Service that will deliver a long-term return on capital. This will be done by examining the aspects of corporate culture that make up a good service regime, by analyzing the service delivery techniques, and by exploring the problems of finding out who the customers are, and what they want.*

CHAPTER 2

Understanding the Customers' Needs

'The marketplace looks totally different from
where the customer is standing.'
Feargal Quinn

If customers' needs and expectations are to be met, they must first be understood. Furthermore, it is not enough for the front-line people, service deliverers, sales people and so on, to understand these needs. Everyone in the company must do so. In *Crowning the Customer*, Feargal Quinn tells how he insists that every one of his executives does the household shopping at least once a month. One of his executives said: 'There's no need for me to do that. My wife does the shopping, and she tells me exactly what is going on.' As Quinn says 'He had missed the point. The customer perspective is something you have to experience for yourself. When you've stood in a queue for four minutes at a time when you're anxious to get away and do something else, you discover how long four minutes can be.' (Quinn 1990, 40)

This chapter describes how customers' needs can be identified and analyzed, and how they can be turned into a service brief. The service brief is the basis on which the other documents and plans of the quality system, specifically the service specification, the service delivery specification and the quality control specification, are built.

Understanding customers and clients

Surprisingly little is known about consumer behaviour. Everyone buys services and products all the time, yet we really do not know what is happening at the moment of truth. We have many models, but little understanding of what makes one service or shop preferable to another, or one advertisement more effective than another. Eighty years ago Gordon Selfridge the store-owner said that 'half of what I spend on advertising is wasted, the problem is I don't know which half'. This is broadly true for modern advertisers. However, it is possible to identify benchmarks by which customer/client behaviour can be analyzed.

Understanding customer activity starts with an understanding of seven key characteristics of buying behaviour. Each of these seven keys affects, more or less, every service purchase. We have traditionally regarded the commissioning of services as a different kind of activity from the buying of manufactured goods. In fact, whether we talk about clients or customers, the same underlying generalizations apply. The purchaser of banking, legal or architectural services is no different in this respect than the same person buying a newspaper, a pair of trousers or the week's groceries.

A good model of your customers' attitudes and motivations is the essential preliminary to a proper service brief. Use these seven characteristics and the sub-points to create a detailed model of your customers' service needs. At many points you will probably not have enough information to make a quantified assessment of the importance of one characteristic above another. You will, however, have reached the first stage of understanding: to discover what you do not know.

1. Consumer behaviour is motivated: People do things for reasons. The reasons may be complex, elusive or buried in psychological motivation, but they are there. In a competitive society, the original practical motive for the purchase of a basic service of law, insurance, accountancy etc. is clear enough, but why do clients go to a small firm rather than a large one or vice versa? Why prefer risky or prudent investment strategies, or go to an established department store rather than the newest boutique?

Some considerations that develop the theory of motivation are:

- clients are driven by tension reduction — a want or a need creates tension until it is satisfied. The failure to satisfy the need aggravates the tension; the satisfaction of the need merely restores equilibrium. This partly explains why complaints are so much more frequent and vehement than compliments. The clients' values will not be understood unless the tension is understood.
- clients estimate the expected value of the service — the greater the tension to be relieved and the more certain it is that the service will in fact relieve the tension, the higher the value and therefore the stronger the purchasing motivation. Barristers, doctors and illicit drug dealers derive their extremely high earnings from being the only people likely to relieve particular high-tension needs.
- motivations are multi-directional, complex and often hidden — individual motivations vary both in energy and direction. Thus the motivation to write a will is less in a healthy person of 25 than in a 65 year old; on the other hand, if the 65 year old was also sick, the motivation might be strong, but

CASE STUDY: HOW TO COMPARE THE QUALITY OF COMPLEX PRODUCTS

1. Each complex product (such as a car, a computer program or a house) has a number of quality characteristics, and each characteristic can be further analysed into attributes. Ideally each attribute should be measurable. The whole analysis turns a complex product into a series of evaluations, item by item.
2. The problem is then how to turn this set of good/bad scores into an overall summary evaluation of the whole product. For instance, if one product scores brilliantly well on certain attributes and badly on others, how is it to be compared with a product that scores moderately well on all?
3. In principle there are two ways of combining measures for individual properties to indicators of overall quality characteristics:
   - if each product property is of equal relevance, the scores from each can simply be added.
   - if each property is not of equal relevance, a weighting has to be given to each.

The possibility of interdependency must be addressed. For instance,

- a fast cheap car is unlikely also to be safe.
- strength and elegance are often difficult to combine.
- instructions for a complex task are difficult to make easy to read.

4. To evaluate this, list all the attributes in polar form, with the positive aspect of the attribute on the left and the negative on the right (see below).

   Since this is a comparative measure not an absolute one, the product/service under test must be compared for each value either with a concept of market average, with a product, or with an ideal or perhaps a minimum requirement.
5. Each product is scored according to how its attributes achieve on a polar score which ranges from + 3 (excellent) to – 3 (very bad). The result is an attribute profile. When compared with other products, the profile indicates how high and low scores correlate.

ATTRIBUTE PROFILE COMPARING TWO CARS

Key: Car A = a a a Minimum requirement: + + + Car B: b b b

Attribute Polar score

+3 +2 +1 0 −1 −2 −3

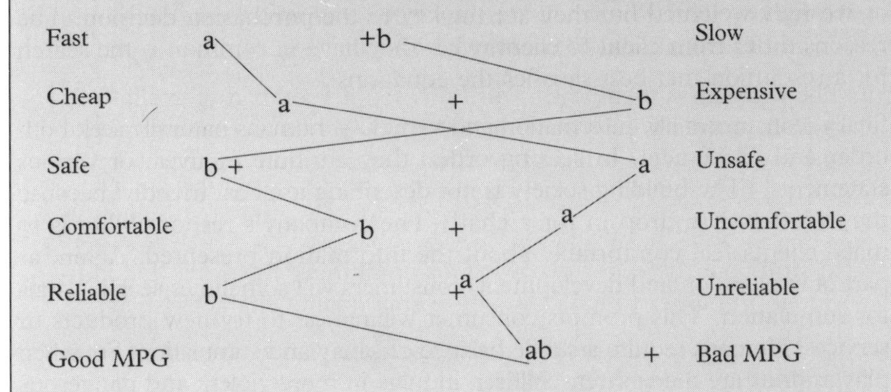

In the example, it is clear that Car A scores better than minimum requirements on speed and price, but seriously underperforms on safety; it is a grade less comfortable than it should be, and its MPG is also too low.

Car B on the other hand scores well on speed, safety, comfort and reliability, and has adequate MPG; perhaps not surprisingly it costs considerably more than was hoped for.

The buyer now has to look for a car which sacrifices some reliability, comfort and MPG for a cheaper price.

strongly negative, as the will would remind the patient of death. Rationality and emotion grapple with each other tragically in this circumstance. Professional buyers are equally driven by complex motivations: personal ambitions, trade friendships, working styles, even ease of parking and access may all come into the purchase calculation, when in theory, service, quality and price are all that are considered.

- consumers have different reasons — all client or consumer behaviour can be said to be motivated by reasons. These reasons may seem eccentric or wrong or wrongly weighted but they are the key to the purchasing decision. The reasons differ from client to client; what they have in common is the search for a resolution that best satisfies the equations.
- finally, consumers like information and variety — humans naturally seek both order and excitement. In seeking order, they attribute motives for various statements. (The building society is not describing itself as 'friendly' because they want you to drop in for a chat!) The company's responsibility is to make clients feel comfortable about the information presented. Again, as part of exploration and development, consumers will sometimes seek tensions for stimulation. This prompts consumer willingness to try new products or services. Humans require a subtle balance of safety and stimulation. So, when playgrounds are made safer, children indulge in more violent and dangerous games. This tension management is an important guide to the rate and style of new service product introduction.

2. **Buying is not just one activity:** From thinking about a service to assessing the cost/benefit after the event, the client goes through a wide range of actions. These include:

- thinking about needs.
- assessing promotional materials.
- asking and receiving advice.
- evaluating alternative ways of meeting needs.
- experiencing the service environment.
- describing needs to service providers.
- assessing price/benefit.
- monitoring and reacting to service provision.
- arranging payment.
- judging the effectiveness of the service.

3. **Consumer behaviour is a process:** Modern studies of client behaviour see the completed transaction as the result of a process. Each of the activities referred to above (complete with its own set of motivations) pushes the client towards or away from the transaction. During this process, the consumer makes four types of decision:

- how much he or she is prepared to spend.
- choice of outlet or provider.
- whether to purchase.
- service specification.

The process divides naturally into three stages: pre-purchase, commitment and post-commitment. In the first stage the two key operations are the identification of the need/tension, and the information gathering. The company's marketing and promotional activity has its most important influence now. The type, presentation and amount of information provided make the difference here between a completed and an abandoned process.

Consumer behaviour and motivation in shops and stores has been widely studied. The psychology of decision-making at the point of sale has also been explored, and theories have evolved of how consumers mix decision rules in coming to a choice. Much less work has been done on how consumers make committing decisions in other service environments such as banks and insurance offices.

In service sales, the purchase decision is a commitment to take part. Once the commitment is made, the service delivery takes place, and the customer immediately experiences the dissonance between expectation and performance. The constituents of this potential source of disappointment are explored in detail in Chapter 10.

4. **Consumer behaviour is a trade-off:** Each consumer wants services that will meet what is perhaps a unique mix of requirements. In practice this does not happen, so the consumer is always in the position of having to trade among sets of attributes. If David and Anna go out to buy a car, for instance, they will typically establish a list of requirements, and then see how well the various possible motors score under each. Some purchases are based on excellent performance in one dimension; others say that attributes A, B and C must achieve a minimum level, but given that, the decider is attribute D. Thus David bases his purchase decision solely on excellent reliability; Anna demands basic reliability, appearance and fuel consumption, and decides on a price from models that meet certain standards.

To achieve a better possible fit to the customer/client's needs, the service provider has to establish an increasingly large 'menu' of possibilities; paradoxically this in itself worsens the service by making choice more complicated.

5. **Consumer behaviour involves several roles:** There are three key roles in every purchase: influencer, purchaser, user. Often two or more of these roles will be combined in one person. Equally often, however, they will be spread. Doctors, influenced by drug company research, recommend drugs which are often paid for by the hospital for use by the patient. Golfers, influenced by sports success, buy 'named' putters and clubs, half hoping to improve their own game by sympathetic magic. Children, influenced by television, persuade their parents to buy the latest craze.

The more significant the purchase, the more influences; no one discusses whether to buy a cheese or ham sandwich with the family; buying a car, an insurance policy or a holiday is different. In these cases as many as four sets of influence might be brought to bear. Most influential are friends, and more generally the potential buyer's peer group; then family; then media generally; then advertising and sales people. Notably, friends and family are not necessarily seen as more expert than advertisers, quite the contrary in fact. They are however seen as considerably more trustworthy and empathetic.

6. **Consumer behaviour is influenced by external forces**: These can conveniently be divided into four categories:

- social and cultural influences — these are some of the deepest prejudices in society such as the relations between the sexes, attitudes to work and play, and class and religious prejudices.
- family influence — particularly strong for many service purchases such as housing, professional services, travel, restaurants and household services.
- life style and situational influences — class, size and source of income and age are three key factors here. Other 'psychographics' categories are more controversial. One early attempt tried to identify people by their response to new products. This divided the population into innovators (2.5 per cent), early adopters (13.5 per cent), majority (68 per cent) and laggards (16 per cent). Another analysis places people in a two-dimensional grid according to whether they respect movement or settlement and values or valuables. Thus 'ambitious' (28 per cent of Irish people) have a high regard for movement and valuables; the so-called 'notables', ultra conservative supporters of traditional ways and ethical codes (23 per cent in Ireland), respect settlement and values. There are almost as many models of this sort as there are researchers in this field; probably each model has most value in the field for which it was evolved.
- advertising and service providers — advertising and promotional activity has most impact, as we have seen, in the pre-commitment stage of the process. It however lays the expectation for what comes afterwards. If I am told the service will be efficient and friendly, then that's what I look for, not long queues and endless forms to fill in. If I'm told the legal service is on a 'no-foal, no-fee' basis, then I do not expect to be charged for 'extras'. If I'm told that 98 per cent of trains arrive on time, I do not want to be told afterwards that the company defines 'on time' as anything that arrives within ten minutes of due time.

The customer comes to the service provider with a set of expectations. In the worst (but frequent) case, the front-line service provider may have only the feeblest idea what these are. To provide the best service, the service provider must be supplied with knowledge both of the product and of the customer.

Studies suggest that empathy with the customer is the more important of these two. Time and money spent educating your customer can be important. This technique has worked particularly well in the vintage wine business, with a constant activity of tastings, chateaux visits and catalogue information available to stimulate the customers.

7. **Consumer behaviour differs for different people:** The model of customer or client behaviour will tend for simplicity to centre on one or more 'average' types. In this complex arena, of course, no one is average. The model must allow for this. At the simplest physical level, perhaps a quarter of the population are short-sighted, unfit, or mildly handicapped; one in twenty suffers from more severe physical or mental handicap.

Similarly, in a retail environment, sales people selling anything from shares to shoe laces will encounter opinionated, suspicious, gullible, enthusiastic, knowledgeable, procrastinating, silent, methodical, timid, independent, conservative, greedy and dishonest types. The effective service response to each of these is different.

Research techniques

The framework explored above provides the basis for an understanding of the complex phenomenon of purchasing activity. It poses the questions. In order to answer the questions, there is a battery of research techniques that may be adopted.

Studying other companies and other industries: This involves actively watching to see what your rivals are doing and analyzing the way they run their business. This does not mean merely picking up the latest trade gossip. It means continually contrasting and comparing, seeking the best solution to the common problem. This is after all what the customers, theirs and yours, are doing all the time. It also involves actively looking at developments in similar businesses for neat solutions to problems or in different businesses but with common strands.

Researching intermediate customers: Dealers, agents and distributors often have their own detailed insight into end-users' needs.

Conducting key-client studies: It is likely that your customers are distributed according to the Pareto distributions: this means that 20 per cent (more or less) of your customers provide 80 per cent (more or less) of your revenue. If these heavy users can be separately identified, explore their value chain, their needs and requirements. This technique operates to an extent in the Frequent Traveller Clubs that airlines have established, and in the better interest rates that banks allow to large deposits. The identifying of special customer groups however is typically done more with an eye to ensuring they stay with the company than as a strategic information resource.

CASE STUDY: HOW THE BANK OF IRELAND ORGANISES CUSTOMER FEEDBACK

Step 1 (6 weeks ahead of Roundtable)
The branch manager contacts the facilitator and agrees the issues that are to be covered at the Roundtable.

Step 2 (5 weeks ahead)
The branch selects the customers it wishes to invite. The basis for selection is determined by the topics to be discussed.

Step 3 (4 weeks ahead)
The venue and all arrangements with regard to meals, etc., are finalised.

Step 4 (3 weeks ahead)
Letters of invitation incorporating detailed information on what a Roundtable involves are issued.

Step 5 (2 weeks ahead)
The facilitator and the branch hold a final briefing meeting.

Step 6 (1 week ahead)
The facilitator ensures that he/she is fully briefed with regard to any specific products or services that may need to be discussed at the meeting. If a particularly complex or technical product or service is involved, the facilitator can request the support of a specialist adviser at the Roundtable to help clarify issues if they arise.

Step 7 The Roundtable normally lasts for two hours. The location is always outside the branch and the time of day at which it is held varies. The branch does not participate in the Roundtable, as experience has shown that customers comment more freely on both the positive and negative aspects of the Bank's service when the session is conducted solely by the facilitator — who is not a member of the staff of the branch. A representative of the branch, however, does attend to greet the customers when they arrive or alternatively to thank them at the end of the session. The facilitator's brief is to create a relaxed atmosphere in which customers feel free to raise issues. A formal agenda is not set, but the facilitator must strive to ensure that issues designated by the branch are discussed.

Step 8 (1 week after the Roundtable)
The branch writes to the customers thanking them for their participation and undertakes to advise them of any action taken as a result of their comments.

Step 9 (2 weeks after the Roundtable)
The facilitator prepares and issues a report covering all the topics which were discussed. This report will objectively record what customers have said and will be divided into two sections, one dealing with specific branch issues and the other with general bank issues.

Step 10 The branch reviews the report and determines what action needs to be taken by their Service Action Team. Copies of the report are also provided to the relevant regional office and the Service Quality Unit. This enables issues which have bank-wide implications to be highlighted and addressed.

Source: Bank of Ireland

Customer panels: It may not be practical to identify the 20 per cent heavy user group. In that case a customer panel, consisting of a cross-section of customers, may be appropriate. This is a small group of customers selected at random, who meet three or four times a year to provide the company with responses to innovations, difficulties and opportunities. About half the companies interviewed for the Humble Report on Service Management in Ireland reported that they used some form of customer clinic.

Response cards: In environments such as hotels, where the customers are transitory and irregular, it may be better to explore satisfaction transaction by transaction. This is usually done by a response card, which the customer is asked to complete at the end of the journey/stay. Comment cards were used by a third of companies in the Humble survey, though they were not regarded highly. Only 8 per cent reported that they found them an effective technique for gaining insights into customer needs.

Other market research techniques: The techniques described above can be set in motion and analyzed by the individual company. Some research techniques are probably best left to market research specialists. Among these are:

- quantitative research, which attempts to identify hard numerical data such as market shares.
- random or quota surveys, as used in political polls. In this technique a representative panel of clients are asked various questions, and the results analyzed. These can be organized on an omnibus basis, where companies can ask one or two questions only for moderate outlay, or a tailor-made research project.

Other techniques which have been evolved to explore consumers' reactions to products include in-depth psychological testing and projective analysis.

Supplier obligations

The service brief must contain an explicit statement of the service company's obligations. These fall into two categories: self- or market-imposed, and legally imposed. The self-imposed obligations, such as warranties and advertising promises arise from a recognition of the needs of the customers. The legally imposed obligations arise from the law. There are an increasing number of laws regulating the supply of services and goods.

The Office of Consumer Affairs and Fair Trade has a 'not entirely comprehensive' list of 110 different laws, statutory regulations, EC directives and similar legal regulations controlling activities which fall into its remit. These include laws relating to restrictive practices, product liability, consumer information, prices, misleading advertising, product safety and standards, food and textile labelling and the sale of goods generally.

When the self-imposed and legally imposed obligations have been identified, they must be addressed. The service brief should contain explicit reference to the particular obligations, and the service specification should make it clear how these obligations are to be met.

THE SERVICE QUALITY LOOP

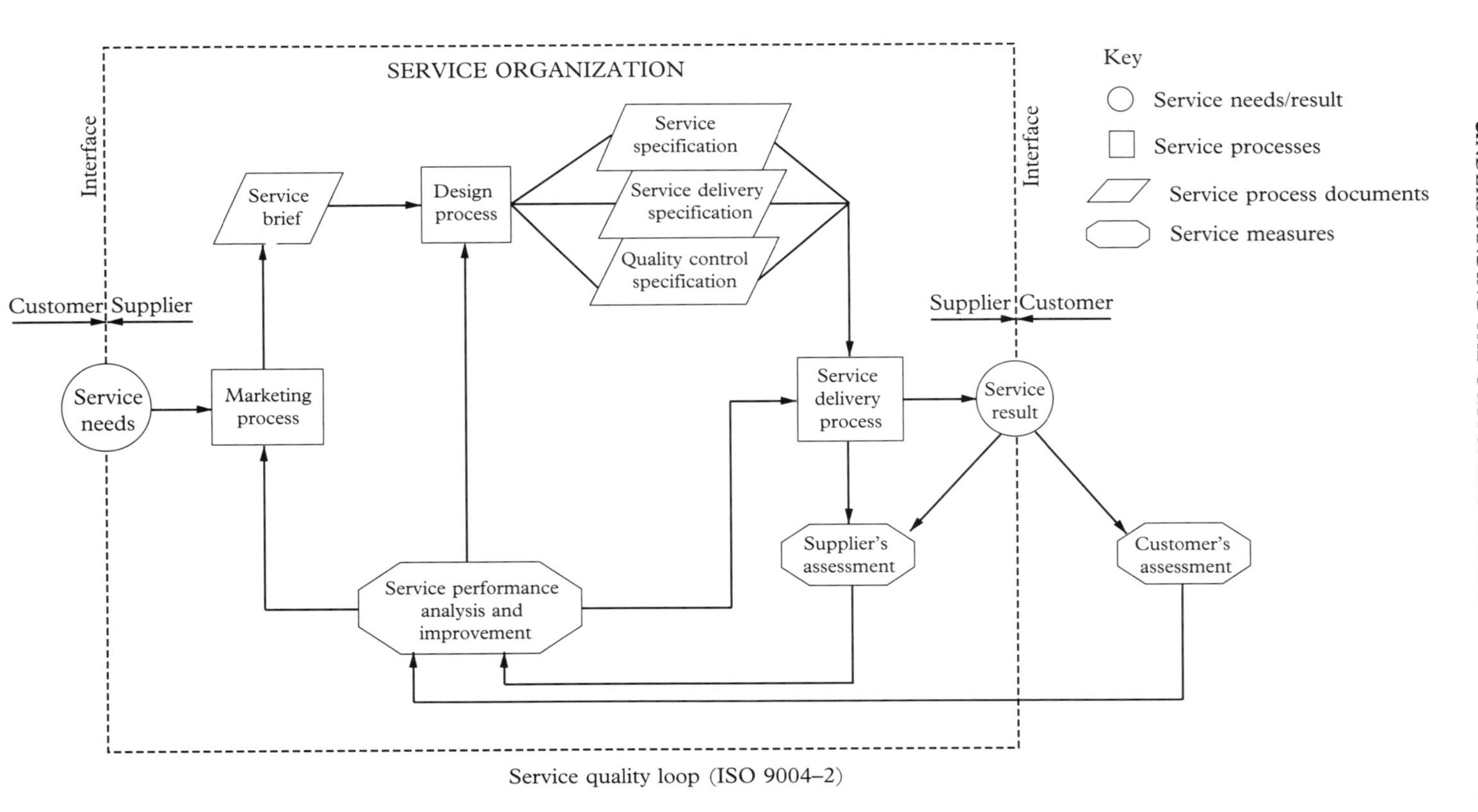

Service quality loop (ISO 9004–2)

SERVICE BRIEF

As we have seen, the ISO 9004–2 Guidelines for Services distinguish four essential and separate written elements in the preparation of a quality service. These are:

The Service Brief, which defines the customers' needs and the related service organization's capabilities and obligations. These are translated into a set of requirements which form the basis for the design of a service. This is the equivalent of the product brief in manufacturing.

The Service Specification, which describes the service which will be provided to meet the needs outlined in the service brief, including the sequence of activities and the key characteristics that affect service performance and evaluation. This is the equivalent of the quality plan in manufacturing.

The Service Delivery Specification, which contains the standard operating procedures describing the methods and measurements to be used in providing the service. This is the equivalent of the set of quality procedures.

The Quality Control Specification, which details the checks and controls that ensure that the operation consistently meets the required standard.

In simple terms, the brief identifies the need the company proposes to meet by providing a service; the specification describes how the need will be met, and what measurements will be used to assess how well the need is met; the delivery specification details exactly the procedures to be used to achieve the standards required; and the quality standard describes the techniques used to assure ourselves that we have done what we wanted to do. This important distinction makes clear the degree of documented detail which the ISO standard requires of service quality management.

In evolving a service brief, the organization must consider the following eight categories of information. All of these elements must have been considered in establishing the service brief. Explicit documentation should be available to prove for instance that competitor activities have been examined and lessons drawn therefrom.

The eight categories identified in ISO 9004–2 (6.1.1) are:

- the establishment of customer needs and expectations relevant to the service offered (e.g. consumer tastes, grade of service and reliability expected, unstated expectations or biases held by customers).
- the investigation of complementary services, and how they might affect the kind and quality of the service supplied.
- how competing activities and performances might affect the service.
- legislation (e.g. health, safety and environmental) and relevant national and international standards and codes.

- understandings of customer requirements and service data that have been collected in-house (relevant summaries of the analyzed data should be communicated to the design and service delivery personnel).
- information from the service organization functions affected to confirm their commitment and ability to meet service quality requirements. It is important to confirm the operations capabilities.
- research to examine changing market needs, new technology and the effect of competition.
- the lessons drawn from quality assurance systems, and quality control results.

These eight sources of information are analyzed and condensed into the service brief. This consists of the following elements:

1. Statement of customers' needs, based on the information gained from market research, from in-house understandings of customer needs and from research into changing needs.
2. Related service operations capabilities, based on information gained from the service providers and from quality assurance results.
3. Statement of the instructions for the operational system. This derives firstly from the first two above, secondly, from the understandings gained of the activities of complementary and competing firms, and thirdly from the legally imposed obligations.

The statements in the service brief are not operating procedures or detailed regulations. They are intended to provide the covering justification and control for whatever operating practices (in the service delivery specification) may be necessary. For instance, if the service brief requires the customer to be supplied with a boiled egg for breakfast, the service brief will state that fact, but will not lay down rules as to how the egg should be boiled or presented.

The instructions should be divided into two specific groups. The first group should explicitly consider how the the four structures that govern the operation of the service are affected. These four groups are:

- management responsibility (see ISO 9004–2 5.2).
- quality system structure (see ISO 9004–2 5.3).
- personnel and material resources (see ISO 9004–2 5.4).
- interface with customers (see ISO 9004–2 5.5).

The second group of statements will cover the operational elements of the system, and how these are to be structured so as to meet the identified needs of the customers. The four quality system operational elements whose activities should be delineated are:

- marketing (see ISO 9004–2 6.1).
- design (see ISO 9004–2 6.2).
- service delivery (see ISO 9004–2 6.3).
- service performance analysis and improvement (see ISO 9004–2 6.4).

A full service brief therefore details what the customers want, what the organization can supply, and how the four structures and the four operational elements are to be combined to provide this service. The next stage is to convert the instructions into specifications for the service and its delivery and control. This stage is discussed in later chapters.

ACTION

- **create a model of your customers' attitudes and motivations, using the seven characteristics.**
- **establish a direct feedback system between your customers and management.**
- **identify your obligations as a supplier.**
- **draft the service brief based on the eight sources of information identified, covering the four structures and the four operational elements.**

SUMMARY

The key to good customer service is an informed understanding of the customers' needs. This can be done by using the seven-point model of customer motivation presented here. Market research and direct feedback techniques are also useful. In evolving the service brief (which formally defines the customers' needs and the supplier's capabilities and obligations) eight sources of information need to be considered. The brief should define precisely how management structures and operational elements are affected by customer requirements.

CHAPTER 3

Becoming Customer Driven

'By customer driven I mean a company where *all* the key decisions are based on an over-riding wish to service the customer better. A company where everyone in it sees the customer as their *only* business.'
Feargal Quinn *Crowning the Customer*

Understanding the customer or client's requirements in detail is hard work, as we have seen. Unfortunately it is only the start of the Total Service Quality quest. The next stage is to instil in all the staff, from the Chief Executive to the night-watchman, a common understanding of the customers' expectations, and a common motivation to meet those expectations. As D.K. Denton puts it in *Quality Service*, in many organizations this will require a change in culture: management 'must adopt a corporate philosophy of quality service by changing the cultural and perceptional attitudes of the organization'. (Denton 1989)

This is called client or customer focus.

Who is the customer?

Most management discussions proceed on the basis that the customer is 'out there', the 'them' in the 'them-and-us' set of opposites. In the direct sense that the customer out there pays the bills, that is true. However this attitude can distance in-house workers from the need for quality. It is too easy, particularly in large organizations, for the pressures and values of the organization to take on a life of their own, regardless of the customers' needs. Too many workers feel 'if only the customers would leave me alone, I could get on with my work!'

As a result, many quality thinkers prefer to think of the customer in this sense as simply the last in a long train of customers. Everybody has a customer, says Richard Schonberger, at the next process, where the work goes next, whether this is inside or outside the firm (Schonberger 1990). As far as the final customer is concerned, the one who pays for the whole chain at the final moment of truth, it doesn't matter whether all or none of the service was created at the point of sale, as long as it meets the need. The fashion industry, where high fashion dresses selling in grand city centre shops for thousands of dollars are stitched for pennies in Asian sweatshops, has demonstrated this vividly.

Each stage in what might be a long process adds some value to the product or service, so that the final moment of truth is as it should be. When McDonald's looked into ways of improving their product, they not only researched methods of cooking french fries, but went right back to the potato growers. They fixed on a particular type of potato (No. 1 Idaho Russet), the ideal period of storage after harvesting (three weeks), even the solids content (21 per cent). They sent 'young clean-cut McDonald's field men armed with hydrometers' to test the product. This was a startling experience for the potato wholesalers whose customers had never before turned up to test specific gravity, or anything else (Love 1988).

This way of thinking about the company sees it as a flow of products (goods or services) through to the final customer. Each stage in the flow expresses its needs to the supplier, provides feedback and so on; each stage is responsible for the quality of the product being passed on. This throws a new light on the strategic make/buy decision. As far as the customer is concerned, there is no intrinsic difference between elements made and provided on the premises, and bought-in products. This is purely a matter of formal organization and preference.

To establish formal, written customer/supplier relations between internal processes may seem overly elaborate. Yet quality lies in the details. If a product goes wrong at any stage, it cannot be made right later; or, if it can be put right, it is always cheaper and better to stop mistakes progressing downstream than to attempt to correct at the end. The internal customer concept has another advantage: it is a way of addressing the problem of skilled workers taking pride and pleasure in the challenge of making the best of bad inputs. Craft workers have usually been trained to make the best of poor quality supplies. Bad inputs provide a challenge; unfortunately with the best will in the world, less can be made of bad inputs than good. However skilled the worker is, quality will suffer. Workers, however, can perceive a threat to their position and job satisfaction in, as they may see it, de-skilling them by attempting to provide perfect supplies.

The chains ahead of the final customer in service industry environments tend to be short. There is therefore usually less scope for establishing elaborate internal customer/supplier protocols. Another implication of the chain-of-customers concept applies most directly to service organizations. Operations should be focused not as is usual by process or product, but by customer. From the organization's point of view it may be efficient to organize a sandwich bar by product counters — meat in one place, milk products (near the fridge) in another, and so on. On the other hand when I go into a sandwich bar, I do not want to buy sandwiches at one counter, apples at another and milkshakes at a third. To provide a customer-focused service, I should be able to go to one service point for my whole meal.

Schonberger takes this idea further by proposing that as far as possible workers should be organized into crews or cells dedicated to one customer-type. This idea gains strength from the understanding that in most organizations the difficult element is not the processing of the work, but the controlling, managing and targeting of it. One should therefore organize the company to minimize the management task, even if this puts more burden on to the operational side (Schonberger 1990).

He suggests organizing groups of workers by client needs first; large scale customers, small scale customers, overseas, corporate, whichever set of requirements makes a coherent package. A group might for instance include sales people, customer service, product service, production chasers, invoice and accounts people etc. They become, as a group, specialists in the needs of a particular set of customers. Information about the customers passes easily around the group, special systems and methods are developed, and a focus on the customers' needs as a whole can be achieved.

This may not be practical in every service organization. The rule of thumb however is to consider the three focus areas in this order:

- customer focus.
- product focus.
- process focus.

Despite appearances in some organizations, the company is not established to make its own internal controls as efficient as possible. The company is set up to deliver a certain product/service mix to the customer. Putting customer focus first ensures that the internal requirements and pressures are not allowed to defeat that ultimate objective.

The moment of truth

Before a company can achieve customer focus, it must be clear what it is focusing on. In Scandinavian Airlines, chairman Jan Carlzon dramatized customer focus by his concept of the moment of truth. 'If you think for a moment', he said, 'you quickly realize that SAS or any other airline *is* the contact between one customer in the market and one SAS employee working at the front line. And when this contact appears, then SAS exists. These are the moments of truth in which we show whether we are a good airline or a bad airline.'

The same concept applies to any service company. The value it creates is in the event of the moment of contact. This may not be a personal contact, though it usually will be. American service quality consultant K. Albrecht describes the point thus in *At America's Service*. 'When a customer sees an advertisement for your business, that's a moment of truth; it creates an impression. Driving by your facility is, for the customer, a moment of truth. Entering a parking lot, walking into a lobby and getting an impression of the place, receiving a bill or statement in the mail, listening to a recorded voice on the telephone, getting a package home and opening it, all of these are events that lead (a customer) to an impression of your service.' (Albrecht 1988)

Two points evolve from this:

- firstly, the moment of truth (which as Carlzon says, *is* the company) exists only as the customer experiences it here and now. No customers, no moments of truth, no company. A spotlessly clean hotel bedroom may as well be filthy if there are no guests to experience the cleanliness.
- secondly, the product is made up of the accumulation of the value created for the customer in these moments of truth.

This accumulation of value does not refer to one transaction only; a key concern of the Total Service Quality strategy is to see customers as providing a life-time stream of transactions. Of course not all moments of truth have the same weight. If we have to choose, most of us would prefer a travel agent to provide fast, accurate information and bookings to a free cup of coffee and a nice smile without these things. It is a key part of defining the service to evaluate correctly the customers' value system in this matter.

In most service industries the whole service experience is made up of many moments of truth. A specialist writer on hotels calculated that 'in a 300-room property running 75 per cent occupancy, with a 2.2 night average stay, it's 37,321 check-ins and the same number of check-outs, it's 73,125 rooms to clean every year, more than 150,000 meals to serve and at least 40,000 morning wake-up calls that must be delivered precisely on time in a friendly, warm manner. It is also the caring for a guest that becomes unexpectedly ill and protecting those who insist, despite cautions, on smoking in bed. It is the safety, security, health, and hospitality day after day after day.' (Hall 1990)

The mission statement

It is up to the company leadership to make it clear to all employees (and especially those in the front line) what the company hopes to achieve in these moments of truth. This is often summed up in a Mission Statement. A mission statement is not the same as a statement of company values, or even as a company policy, nor is it some form of internal advertising. It is a concise and memorable description of the values the firm is planning to deliver to the customer. It is the answer to the customer's question: what can your organization do for me?

In advertising jargon this is sometimes called the Unique Selling Proposition (USP). The USP consists of three elements:

- a definite proposition or benefit.
- a unique benefit, one that no one else is offering.
- a saleable benefit, since there is no point in uniquely offering something no one wants.

CASE STUDY: MONITORING CUSTOMER FEEDBACK TO THE ESB CASH OFFICE

Cash Office — Year to Date

Quarter Ended	Number of Responses	Quality of Service			Helpful?		**Queues?				
								Sometimes		Frequently	
		Friendly	Indifferent	Unfriendly	Yes	No	Never	>4 Mins	<4 Mins	>4 Mins	<4 Mins
Dec '90	1,362	90%	9%	1%	97%	3%	58%	15%	24%	2%	1%
Mar '91	1,021	91%	8%	1%	97%	3%	57%	14%	26%	2%	1%
Jun '91	1,124	89%	10%	1%	97%	3%	58%	16%	22%	4%	–
Sep '91	1,076	90%	9%	1%	96%	4%	57%	20%	19%	1%	3%
Overall	4,583	90%	9%	1%	97%	3%	57%	16%	23%	2%	2%

Source: ESB

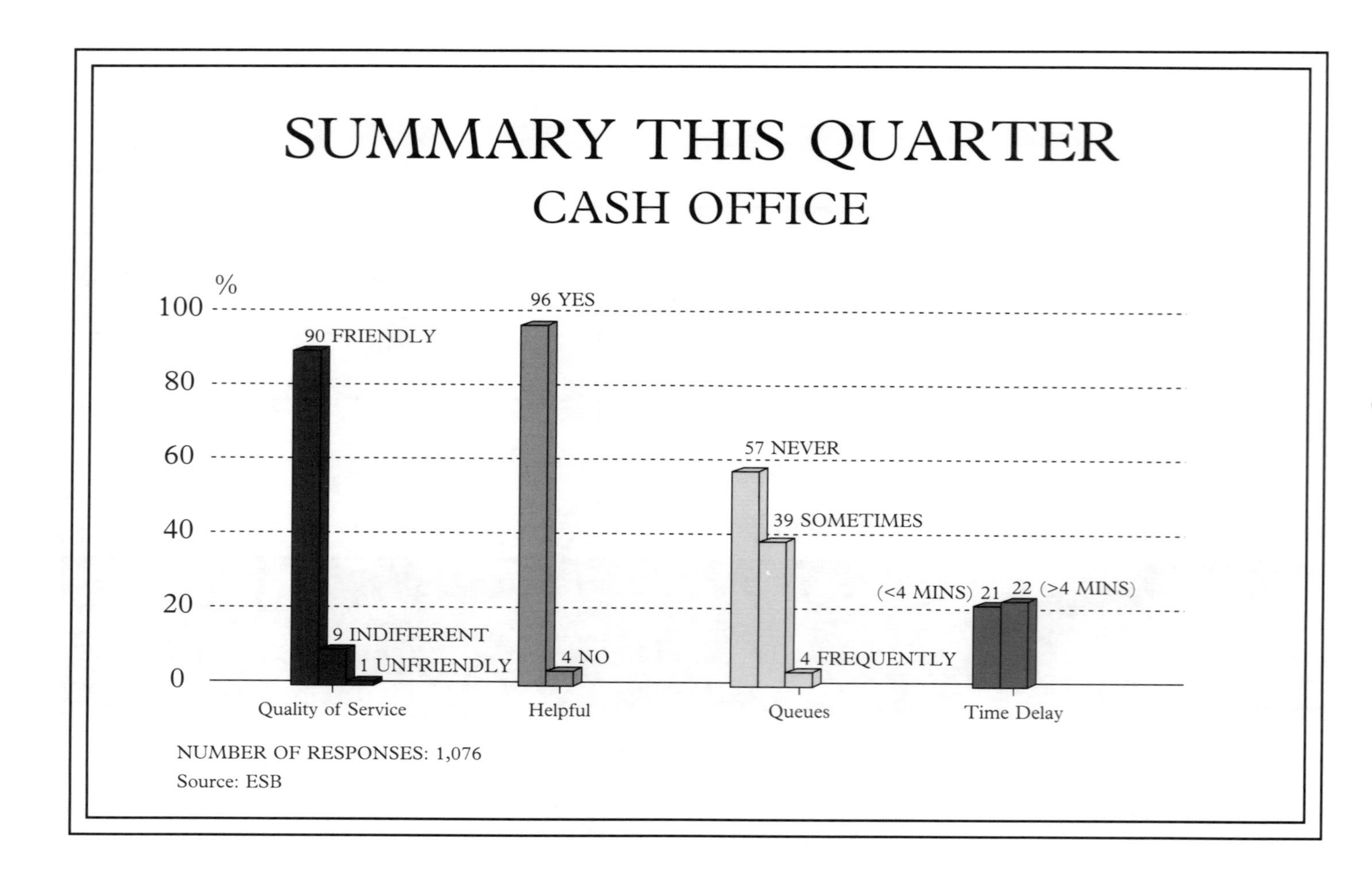
SUMMARY THIS QUARTER
CASH OFFICE
%
100
80
60
40
20
0
90 FRIENDLY
9 INDIFFERENT
1 UNFRIENDLY
96 YES
4 NO
57 NEVER
39 SOMETIMES
4 FREQUENTLY
(<4 MINS) 21
22 (>4 MINS)
Quality of Service
Helpful
Queues
Time Delay
NUMBER OF RESPONSES: 1,076
Source: ESB

The key lies in the uniqueness. Strictly speaking, the service you are identifying as your USP may be offered by other companies. You select it as important because it identifies a touchstone of your business. Thus all booksellers will, sometimes reluctantly, order books specially; if one announces A Special Book Ordering Service, and gears the business round that, they have identified a USP. Equally all lawyers operate on a 'no-foal, no-fee' basis from time to time; if one announces this as an identifier, it becomes the USP.

The uniqueness of your service attracts new customers and holds existing ones. But what exactly is the USP? Are you simply larger, or better known, or related to more people, or nearer, or open longer hours than the rivals? If so, these seem flimsy enough as USPs; perhaps a little more research might be called for into exactly what customers need. Tom Peters, in *Thriving on Chaos*, maintains that the successful firm in the 1990s will be tightly focused on specific niches in the marketplace. Especially in the service market, people are looking for high values of reliability, flexibility, empathy and so on, and they are prepared to pay for it. The key to successful marketing is to find a niche and become the best in that niche. The USP is the expression of the uniqueness.

A mission statement can become the driving force of client focus in the company. Being short and memorable, it is easily retained by everyone, and everyone can relate their job to the mission statement. Everyone can see precisely what it is the company proposes to do for the customers, and what's more they can see exactly how what they do furthers that end. If there is not a fairly clear connection between what they are doing and the customers, there may well be a question mark over the job they are doing.

The ideal mission statement is:

- short — less than 25 words.
- memorable — simple words, elegantly connected.
- precise — an exact expression of what attracts new customers and holds existing ones.
- not full of vague aspirations such as 'making the world a better place to live in', or 'serving the community'.
- extremely well-known to the workforce — its purpose is defeated if it is not used as a daily touchstone for activities and attitudes.

Examples of successful mission statements are the home-delivery pizza firm's 'Gourmet Pizzas in 30 minutes', another is the computer manual printing company Microprint's statement 'Quality print, just in time'.

Achieving a client-focused company culture

Client focus, particularly for large companies with elaborate in-house routines, can be difficult to achieve. It can require a totally new orientation for many attitudes and practices. In fact, the degree of client focus is one of the ways in which an organisation's 'corporate culture' is assessed. This concept is an extremely important one for total quality management. We all know that some families are very hospitable, some less so; some are house proud, or fond of family jokes and traditions, or very independent of each other; some are disciplined, achievement oriented, or easy going. Each family has its own way of doing things. Each company, likewise, has its own way of doing things, and its own attitudes. This set of attitudes makes up the corporate culture. It is the dominant force for good or evil inside the company.

Consequently, creating and developing an appropriate culture is pre-eminently a task for the very highest level of management. In fact Schein, one of the leading experts in this still hazy field, has suggested that 'the only thing of real importance that leaders do is to create and manage culture'. If the role of leader can be distinguished from administrator or manager it must be because the unique talent of leaders lies in their ability to manage culture (Schein 1985). Certainly, as consultant after consultant has stressed, Total Service Quality will not occur in a company without the active and visible participation of the Chief Executive.

What then *is* a corporate culture? Schein defines it as 'the set of basic assumptions and beliefs that are shared by members of an organization, that operate unconsciously, and that define in a basic "taken-for-granted" fashion an organization's view of itself and its environment'. He goes on to stress that these assumptions and beliefs are '*learned* responses to a group's problems of survival in its external environment and its problems of internal integration'. What has been learned, he notes, can be changed.

Corporate culture can be seen in operation in various ways. Two particular manifestations are extremely common. These are management's obsession with short-term financial results, and the inward looking organisation.

1. Management's obsession with finance: The first and most pervasive obstacle to client focus at top levels of management is the belief that a businessperson's primary concern is with the financial figures. This belief comes from several sources. Firms are assumed, by economists, always to act in such a way as to maximize their profits. (See for instance the undergraduate textbooks that trained many of today's CEOs, such as Lipsey 1963.) Such is the respect in which economics is held that there is an implicit feeling that if a businessman does not put profit first and last in his calculations, he is in some way flying in the face of nature.

This feeling that numbers are the key to business management is reinforced from a different point of view by the accountancy profession. Since the Second World War it has become essential for large and small businesses to have a qualified accountant at the heart of deliberations. These accountants may know little or nothing about what the company does: that is not regarded as important. As one leading Irish accountant put it 'accountancy gives you the great entrée because you understand numbers. Out of that can come a vision of what can be achieved in business . . . unless a company is successful and making profits, then it cannot achieve its other objectives. If it's a design company or a manufacturing company or a retailing company, unless it's making money it cannot do any of those things. The whole thrust of everything is getting the bottom line right, and if your bottom line is right you can give your attention to the strategic planning of the business.' (*The Irish Times* 16 March 1991)

This clearly puts the cart before the horse. The company can only affect the bottom line by good designing or manufacturing or retailing. Good accounting merely keeps the score. Unfortunately many companies in the last twenty years or so have suffered from this attitude; the undoubted professionalism and competence of professional accountants, and their possession of a clear theory of business success has mesmerized many a board.

As it happens, even accountants are beginning to come to this view: as one textbook put it 'firm's accountants . . . have become removed from concerns as to whether the numbers they were objectively and consistently recording held any relevance for describing, motivating and controlling the firm's manufacturing performance'. (Cooper and Kaplan 1991)

The primary focus of everyone's activity should be on the customer and the product; only secondarily, as a check up, should the accounts be consulted. If you have a product that everyone wants, you score well; if not, you score badly. Profit, like happiness, follows from activity, not the other way round. A regular health check is good preventive medicine but a daily visit to the doctor is neurotic. (This criticism of the effect of the financial scorekeeping function on management thinking is not a criticism of the importance of measurement in business. In fact, as Chapter 13 shows, proper measurement of key financial and non-financial variables is essential to proper management.)

2. The inward looking culture: The exaggerated importance of the financial scorekeeping function is not the only way in which the customer is undermined in organizations. In most organizations there is a dense structure of communications, commands, social interactions and emotions. 'The railways,' said a station-master recently, 'are an octopus-shaped town, its tentacles spread all over the country. No matter where we come from, we all belong to the same place.' (*The Irish Times* 23 July 1991). Where, in this town, do the passengers live? Managers and workers communicate frequently with their superiors, their superiors' superiors, with line colleagues, with staff colleagues, with various levels of subordinates, with suppliers, with professional associations — and occasionally customers. Tom Peters recalls a game he plays at his in-company seminars. He asks managers to examine their in-trays and divide the contents into two piles: those relating directly to the customers, and those relating to in-house matters. Inevitably three-quarters of most managers' in-trays concern in-house matters.

The larger the organization, the safer the market, the more prone it is to this dangerous form of self-obsession. In 1965 *The Irish Times*, after an industrial dispute, quoted an airline spokesman as summing up the result: 'What we're doing is in the interests of everybody, bar possibly the consumer.' Organizations with a strong internal culture, such as very large manufacturing companies, hospitals, railway companies and professional firms, are very prone to this.

The excellent companies identified in Peters and Waterman's *In Search of Excellence* make elaborate efforts to evade this culture-boundness. Disney executives spend a week a year 'on-stage' with the public, selling hot dogs and popcorn, or driving monorails. Other excellent companies go to similar lengths.

These are two of the most common maladies of corporate cultures. There are other more subtle symptoms that need to be explored, for instance:

- how far do top management trust and listen to subordinates?
- how far are rules and systems rather than results stressed?
- what are the penalties for making a mistake?
- what kind of feedback systems operate?
- how far are negative or positive motivations used?
- does the work-place operate as a social centre, or is it totally job-centred?
- are goals and corporate objectives set with consultation down the line?
- what type of people get on in the company?
- what are the dominant values? (Is it profit, client service, covering one's back or controlling costs etc.?)
- what importance is given to physical layout and conditions?
- what are the normal attitudes to work?
- do the workers trust the management?

The answers to these questions are part of every management's inheritance, whether from its own actions or from those of its predecessors. The attitudes identified by these answers significantly limit freedom of action. To achieve Total Service Quality necessarily implies a deliberate change of corporate culture. This must be driven, managed, and controlled by management.

The first step is to identify as coolly as possible the current state of the corporate culture. This should probably be done by an outsider, since this is in effect a radical critique of management. In briefing the outsider, it is important to make clear the purpose of the exercise. What is required is a detailed and scaled report identifying potential problems in the introduction of Total Service Quality. Once the state of the culture is identified, the next stage is to prepare the management and then the workforce for the change.

But how can such deep and sensitive feelings be changed? Schein identifies various ways in which a new culture can be grafted on to the old. This process must necessarily be a slow one, by evolution rather than dictat. To take an extreme case: a notice on the company board to the effect that: 'Following the introduction of Total Service Quality, from Monday all workers will trust management' is unlikely to change much if for years management has been legalistic on deals, has introduced major changes which affected earnings without consultation, and constructively dismissed workers. For years that management has been teaching one corporate culture; now the old has to be erased and a new one put in its place.

The kind of change that is possible depends, says Schein, in large measure on the development stage of the organization and the degree to which it is ready to change. The most difficult case is a large, complacent body with highly elaborate systems and a long tradition: typically a railway system, or a public utility. On the other end of the scale is a new company set up with all graduate staff (mostly on their first jobs), in a new fast-moving industry such as computer technology. Each of the strategies follows a standard path:

- identification of attitudes which are no longer effective.
- unfreezing of old attitudes.
- creation of psychological safety for those asked to adopt new attitudes.
- a mechanism to install new attitudes.
- confirmation and consolidation of the new attitudes.

The various mechanisms for change identified by Schein are:

1. Natural evolution: In this case the company builds on what it does well and what works; this is ideal in a stable market.
2. Self-guided evolution: In more dynamic markets, consultants may be necessary to unfreeze the company, to provide self-confidence, to help analyze the problems and opportunities, and generally to guide the company into a new stability with the environment.
3. Managed evolution through hybrids: 'Hybrids' are existing staff members who though they have ingested the existing culture have reacted differently to it. Their personal assumptions are different and more suitable to the new requirements than the existing acculturated managers.
4. Managed revolution by outsiders: If no hybrids are to be found, management may have to introduce outsiders to develop new approaches. A typical example of this is when an expanding company employs professionally trained managers, such as accountants, to provide a different edge to management skills.
5. Planned change: This is similar to self-guided evolution, but in the context of a fully established professional management team rather than a small group of founders.

6. Technological seduction: New technology forces change. The introduction of computers into offices 20 or 30 years ago made shift working among white collar workers respectable, even glamorous; the use of electronic mail can change the whole company's attitude to secretaries and ancillary staff.
7. Change through scandal: A culture will tolerate a surprising amount of abuse, until someone goes too far. Then the explosion causes major revaluations and revisions of practices and attitudes. The most spectacular incidence of this was the complete revision of London's Fleet Street working practices by Rupert Murdoch.
8. Incrementalism: It is said that a frog can be boiled to death without protesting if the incremental changes in water temperature are gradual enough. Psychologists have discovered that many quite radical changes, for instance in tolerance of lighting, can be introduced undetected by a series of small changes (below the Just Noticeable Difference (JND) level). This may of course be a long process.
9. Coercive persuasion: In moments of severe corporate difficulty, there may be no option but to change or leave. Senior management has to control the unfreezing and challenge the old assumptions, if necessarily coercively, to shoe-horn the new ways into the routines. This may be part of a deliberate turnaround policy to rescue an ailing company. In a turnaround situation, some or all of the above techniques may be brought to bear.
10. Destruction and rebirth: The Doomsday scenario. In theory the parts are scattered to be reformed elsewhere as a new operation. In practice each worker learns from the experience, which now forms part of the set of assumptions carried to the next job.

These mechanisms for initiating change are the ways in which a new customer-driven culture can be installed. Throughout the process, senior management must remember the Three C's of Culture Change:

Commitment: Instil a commitment to a common philosophy and purpose, recognising that employee commitment must coincide with individual and collective interests.

Competence: Develop and reward competence in key areas, keeping in mind that you will foster greater competence by focusing on one or two key skills at a time rather than by addressing a host of skills all at once.

Consistency: Be consistent; let actions and words support, not contradict, each other.

(Hickman and Silva 1985)

ACTION

- **identify all the moments of truth in your company's operation.**
- **what are the values the customers seek at each of these moments of truth? Do you know? Do the staff at the sharp end know?**
- **identify the chain of internal customers and the values that must be transmitted up that chain to achieve the satisfactory result at the moment of truth.**
- **start the quest for a mission statement.**
- **what are the positive and negative aspects of your company culture?**

SUMMARY

Quality starts and ends with customers. For quality to work, the workers, especially those at the service delivery point, must have a clear understanding of the customers' needs. The achievement of this requires minds and activities to be focused according to customers' needs. The obstacles to customer focus draw executives' attention to internal matters such as cost structures and internal hierarchical problems, rather than to the customers' needs.

The importance of client focus, especially for service companies, is clear. The ideas outlined in this chapter start the company off on the continuing journey towards that goal. Together the ideas form a powerful armoury. The paradigm shift in organizational self-image is the start. A company should be seen not merely as a device for channelling commands from the highest to the lowest as quickly as possible, but as a support group to those delivering the goods to the customer. Exactly what those goods are is defined first of all by the mission statement, which encapsulates the organization's raison d'être, and the moments of truth concept, which provides a dramatic image of the mission statement in action.

CHAPTER 4

Service and Service Delivery Specification

'Today we have naming of parts. Yesterday
We had daily cleaning. And tomorrow morning
We shall have what to do after firing. But today,
Today we have naming of parts.'

H. Reed *Lessons of War*

Every service and every product encompasses a mixture of service elements, product elements and price. Each has to meet customer expectations. In the case of a simple sandwich bar, fast friendly service can hardly compensate for stale ingredients; conversely, even the best ingredients would not survive slow and surly service. Price can temporarily cover up the cracks in either service or production values, but this is not likely to last long.

In considering the service delivery specifications, management has to decide on the values it gives to the various factors involved in giving a service. To explore the weights and factors, the Marketing Science Institute of Cambridge, Massachusetts financed a research programme which began with hundreds of interviews with customers of several different service sectors. The researchers, Zeithaml, Parasuraman and Berry, eventually identified five key factors in service. Their findings were published in one of the best books on service quality, *Delivering Quality Service, Balancing Customer Perceptions and Expectations*. The five factors were:

1. *Reliability*: defined as 'the ability to perform the promised service dependably and accurately'.
2. *Responsiveness*: defined as 'willingness to help customers and provide prompt service'.

3. *Assurance*: defined as 'knowledge and courtesy of employees and their ability to convey trust and confidence'.
4. *Empathy*: defined as the 'caring individualised attention the firm provides its customers'.
5. *Tangibles*: defined as 'appearance of physical facilities, equipment, personnel and communication methods'.

When these five factors were played back to a sample of nearly 2,000 customers, the researchers were able to put the factors in order of relative importance. Out of 100, reliability scored 32, responsiveness 22, assurance 19, empathy 16 and tangibles 11. In other words, 'appear neat and organized, be responsive, be reassuring. Be empathetic, and most of all be reliable — do what you say you are going to do.' (Zeithaml, Parasuraman, Berry 1990, 27)

This order is intuitively believable. The first requirement must always be reliability: we want ATMs to stay open, department stores to deliver when they say they will, and not several hours later, accountants whose tax returns are accepted by the Revenue, planes to leave and arrive on time. This stress on reliability is simply a reaction to experience. In 1984 research found that 19 out of 20 companies deliver products late (Atkinson 1990). The first step in the development of a quality programme must therefore be to address the *reliability* of the service. No amount of redecorating the service hall, smiling, or calling the customer by name will compensate for incompetence.

The next focus of the service specification must be on *responsiveness*. From time to time things go wrong, or a customer has a special need. The ability of the firm to respond quickly and effectively to this situation must be the second focus of development of the quality programme. To some extent this can be built in to the quality procedures. Eighty per cent of problems are likely to arise from 20 per cent of the customers. Contingency plans to handle these difficulties can be drawn up. However, there will always be the unplanned occurrence; in handling these situations the real quality of the staff comes through. Staff members will have to react immediately to a problem, often in a stressful situation with the customer glowering over the counter.

At this moment of truth, the service giver has to calculate the whole firm's response to perhaps a demanding and awkward customer. Various thoughts flash through the server's mind: the first problem is to protect him or herself. A server who really believes that the boss is serious about client satisfaction (and this is not to be achieved merely by erecting notices saying 'The Customer is Always Right'), will feel confident in making disproportionately generous offers of the firm's resources to put the problem right. Very positive effects for the firm's future image can be achieved by 'over-reacting' in this way to a problem. A culture in which such an offer will be scrutinized and criticized is likely to deliver a very negative reaction to the customer at the moment of truth. The quality of responsiveness in these circumstances is rightly valued by customers: it can only be achieved by a firm, even messianic, commitment to the customers' values by management.

Assurance is the next value that the service planner must build in to the service. This is an important value in many transactions: the famous advertising campaign 'No one ever got the sack for buying IBM', was built round this value. In service quality environments it can be especially important. Many service customers rely heavily on the external assurance factors to judge the quality of the product. Doctors' waiting rooms, heavily furnished and lined perhaps with degrees and certificates, crisp linen table-cloths in restaurants, smiling in-flight cabin service, are all outward signs of inward grace: reassurances that all is well.

No purchaser, whether of widgets or haircuts, likes to feel part of a crowd. We all feel our requirements are slightly different. As a result we value *empathy* highly. A service organization must make a customer feel special, that his or her individual needs are being responded to, to secure their loyalty. However this empathy is not a substitute for, nor even a priority to the values of reliability and responsiveness. One of the great service companies of the last 30 years, McDonald's, has made a virtue of providing a minimum of empathy and individuality; reliability of product and service quality were the values they emphasized from the beginning.

The final key quality dimension, *tangibles*, is the least important. The appearance of physical facilities, uniforms, equipment, personal appearance of staff are relatively easy to get right.

The ISO 9004 Guidelines distinguish the service specification from the service delivery specification by noting that the service specification 'defines the service to be provided, whereas the service delivery specification defines the means and methods used to deliver the service'. The two are obviously closely linked. 'Design of the service specification and the service delivery specification are interdependent and interact throughout the design process.' (*Guide to ISO 9000 Quality Management in the Service Industries*)

The ISO Guidelines require that 'the service specification should include:

- a complete and precise statement of the service to be provided, including service characteristics subject to customer evaluation.
- a standard of acceptability for each service characteristic.

The service delivery specification should include:

- service delivery procedures describing the methods to be used in the service delivery process.
- a standard of acceptability for each service delivery characteristic.
- service delivery resource requirements describing the type and quantity of equipment, facilities, and personnel necessary to fulfil the service specification. Reliance on sub-suppliers for either products or service should also be described.' (*Guide to ISO 9000 Quality Management in the Service Industries* 5.2.3 & 5.2.4). The technical term 'service, or service delivery, characteristic' describes the individual elements that directly affect service performance.

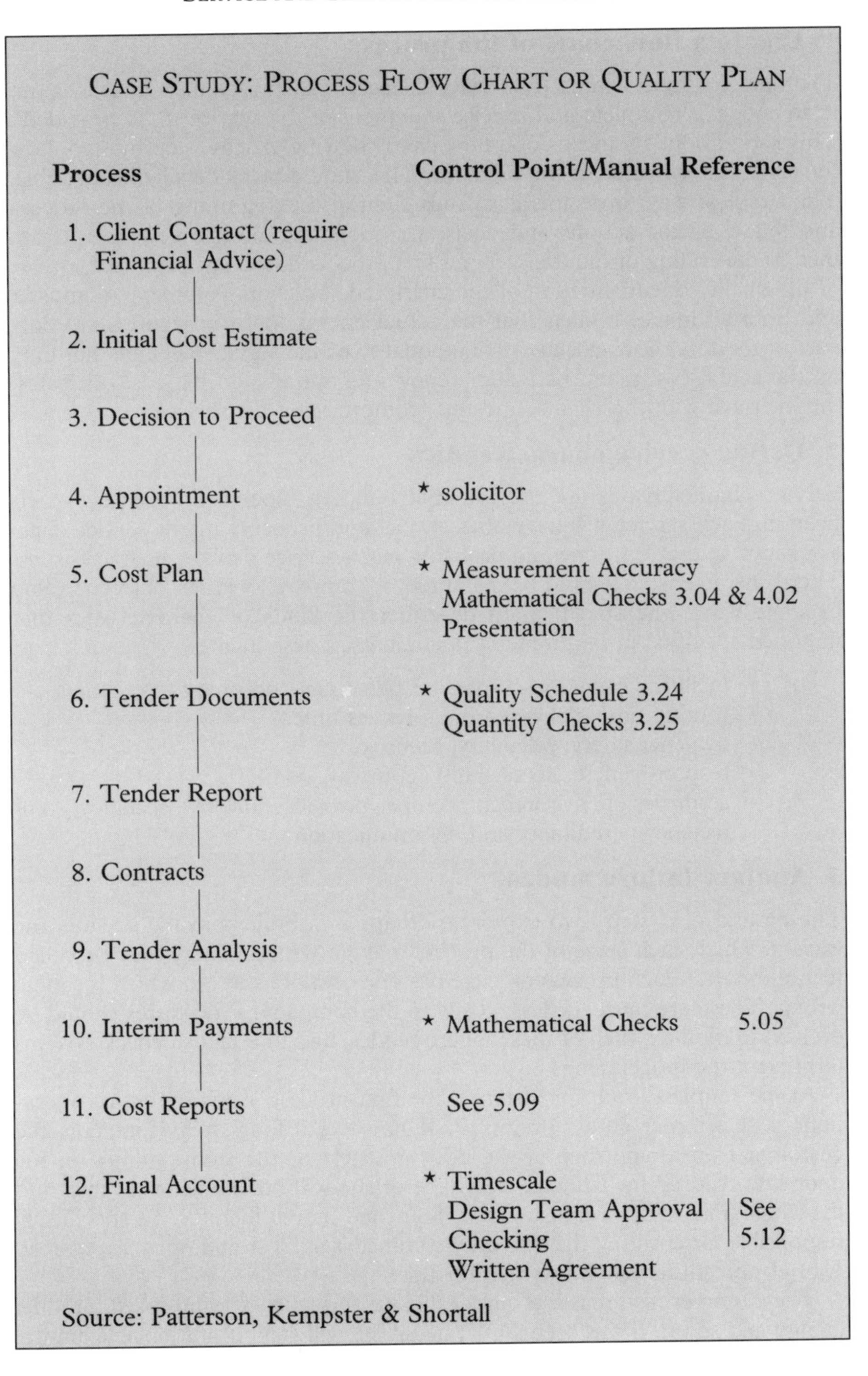

CASE STUDY: PROCESS FLOW CHART OR QUALITY PLAN

Process	Control Point/Manual Reference
1. Client Contact (require Financial Advice)	
2. Initial Cost Estimate	
3. Decision to Proceed	
4. Appointment	* solicitor
5. Cost Plan	* Measurement Accuracy Mathematical Checks 3.04 & 4.02 Presentation
6. Tender Documents	* Quality Schedule 3.24 Quantity Checks 3.25
7. Tender Report	
8. Contracts	
9. Tender Analysis	
10. Interim Payments	* Mathematical Checks 5.05
11. Cost Reports	See 5.09
12. Final Account	* Timescale Design Team Approval | See Checking | 5.12 Written Agreement

Source: Patterson, Kempster & Shortall

1. Create a flow chart of the process

The first step in designing the service delivery system is as the Guidelines put it, to create a 'complete and precise statement of the service to be provided'. This is best done by means of a flow chart showing exactly, step by step, how the service is delivered at the moment. This should be as detailed as possible. The simplest way to do this is to start right at the beginning of the process and follow all the actions and decisions (including alternative routes) from then to the ending of the transaction. The process flow chart produced as part of the quality documentation of the chartered surveyors Patterson, Kempster and Shortall makes it clear that the actual operational transactions are only part of the total flow of activity. The quality of the service depends not only on the actual cost plan, the tender report etc. but also on the cost estimates, the interface with the clients, and the promptness of the service.

2. Define service characteristics

Service characteristics are the minimal building blocks of a service specification. Each service will have at least one, and probably many, service characteristics, and at least one and probably many service delivery characteristics. These characteristics should be measured or compared with an objective value in some way. The specification describes the kinds of characteristics that might be specified in requirement documents. These include:

- facilities, capacity, number of personnel and quantity of stores.
- waiting time, delivery time, process time.
- hygiene, safety, reliability, security.
- responsiveness, accessibility, courtesy, comfort.
- aesthetics of environment, competence, dependability, accuracy, completeness, credibility and communication.

3. Analyze failure modes

The natural next step is to explore, perhaps in a brainstorming session, the ways in which each stage of the process may go wrong. These may be called failure modes. Each process or stage has one or more ways in which it can go wrong. These are more, or less, costly to the company. Procedures should be evolved to monitor each of these failure modes, and to establish check systems to prevent them occurring.

At the simplest level, for instance, the first process in many service operations is 'Customer enters premises'. If however a long queue prevents the customer from doing this, or the door is stuck, or the sign hanging on the door says 'Closed for Lunch', clearly he or she will be unable to do this, and the whole transaction will be aborted. Someone should therefore be specifically responsible for ensuring that the door/entrance is as clear and open as possible. A brief procedure may be written for this.

The customer next makes a quick estimate of queues, serving speed, product quality and so on. If this seems OK, he or she will stay; if not, once again the transaction just does not happen. This at once directs the manager's analysis to the physical appearance of the product or service, and the queue length and speed, which in turn brings up the question of the number of service points. Clearly some system could be evolved to bring in another server when the number of people in a queue exceeds an agreed total.

The seriousness of each failure mode can then be estimated: it is clearly a seriously expensive failure if the customer cannot get inside the shop. A failure to ask if the customer would like something else is obviously less serious, but still loses potential sales. A cost or seriousness factor can be evolved for each failure mode, thus directing the control effort. Using data based on the TARP surveys of repurchasing intentions, a company can create a formula for calculating the cost of lost sales due to specific problems. (E.g. Lost due to problem = (% experiencing problem x % likely not to repurchase) x total sales.) (See J. Goodman 'State of the Art Techniques for Measuring and Monitoring Customer Satisfaction', a paper given to the MCE Conference 'Service: the new competitive edge' in London 1991.)

4. Establish standards of acceptability

Another output of this analysis is the evolution of objective standards of acceptability. These will be of two kinds: quantitative and qualitative. Quantitative standards are those that can be achieved by measuring and counting. Thus quantitative standards could be established in a sandwich bar for the average length of time between entering the shop and leaving it with the sandwiches required; for the average hours since the bread rolls were baked, and the amount of butter, salad and meat included. A typical objective standard is the time spent queuing to give your order. In this kind of business the more people you can serve in the short peak-demand period, the higher your return on capital.

Since customers are likely to take about the same time to serve, the length of queuing time will be in direct proportion to the number in the queue. You might start with the assumption that more than six (say) in the queue means an unacceptable wait. On the other hand most of the customers may not at all mind a longer queue time and a chat; or that a certain proportion of customers are buying large orders and therefore throw the calculations. Whatever the specific circumstances, it is important that the queuing time be monitored as a key strategic area. (The whole question of queues and schedules is discussed in greater detail in Chapter 8.)

Incidentally, there should not be too many measures of acceptability. A few well-understood measures are likely to be much more effective than little measures checking every detail. Because information is so heavily computerized, in the financial sector it is tempting to multiply measures. One banking organization in America produced a Quality Measures Master List including measures such as:

- number of incorrect standard and savings accounts documents as a percentage of all new accounts.
- complaints about missing or wrong statements as a percentage of total statements.
- coding errors per million.
- average customer waiting time.
- dollar errors at end of day.
- letters not answered within two working days.
- calls answered in more than 10 seconds as a percentage of all calls.

In all, the list contained over three hundred measures covering 29 areas of the bank's operation. This density of information is more likely to paralyse action than to stimulate it.

Qualitative standards deal with intangible characteristics. These are often hard to measure precisely. It is always possible however to translate these characteristics into scales. A customer may not be able to quantify how reliable your firm is, but one can always make a judgment about whether a firm is 'very poor, poor, average, satisfactory, excellent' under this and many other characteristics. The numbers ticking each level can then be added to enable regular monitoring to take place. It may be as well when assessing these tick forms to even out the differences between respondents. This can be done by counting only three categories: very poor, excellent and the three middle categories together.

The problem then is to get a stream of these judgments from your customers. In some cases it might be suitable to stimulate participation by, for instance, holding a draw using the returned questionnaires, or giving a discount on the next purchase. Many firms, especially in the hotel business, use comment cards. Other methods used to establish customer opinions are questionnaires, customer 'clinics' (as used for instance by the Bank of Ireland — see case study), market research, salesforce discussions and direct customer/management meetings. The Irish managers who were questioned regarded only the last of these as being largely effective.

The ISO guideline seeks to ensure that, as part of the service quality management system, service companies should have both a detailed analysis of the service delivery process, and the standards of acceptability for each service characteristic. However these matters, though important, are essentially static. The next section deals with the basics of the dynamic process of service design, using the service blueprint technique.

Service design

The importance of good service design is only just being appreciated. As Professor Gummesson, Professor of Service Management at the Service

Research Centre in Sweden points out, 'it takes five years to design a new model of car in Europe. In order to design the car, detailed blueprints, specifications, and prototypes are produced. Special computer programs are used. The car is tested in laboratory settings as well as in natural settings. Large manufacturing companies may employ thousands of designers to develop, improve and customize products. We have yet to hear of service designers. There are specialists in the field such as interior decorators for restaurants. There are computer programmers and systems specialists to help banks and insurance companies. There is however no tradition of service design. There are no CAD programs for services.' (Gummesson 1989)

This is certainly true for Ireland. Creation of the service brief is perceived as a 'creative' or administrative process, not a systematic one. Someone might have a bright idea about design or ambience (for instance, Captain America's) or for a completely new service (for instance, print your own slogan on a tee-shirt), but there is no systematic analysis and development. There is little sense of the *possibilities* that good service design can expose. With a bit of lateral thinking, the ways of handling a service can be multiplied and explored. Statistics tell us that there are as many as 2.6 million ways in which a simple hand of five cards can be dealt in poker: just over a million of these will contain a pair, and 54,912 a three. A few of the hands will be markedly better than these, such as the 624 chances of getting a four, or the 40 chances of getting a straight flush. Service design aims to get at least a four.

The problems are however not so neatly computable as a pack of cards. Whatever the difficulty, the organization should make every effort to acquire a winning hand by exploring options and ideas, and not simply assume that the first ideas you had were right. For instance most of the activity in a sandwich bar is getting the order to the sandwich maker and back again. Having identified this, you may decide that you wished to break the connection between the order taker and the sandwich maker. There are various ways you could do this. Orders could be taken by one set and passed through to the kitchen for processing, like a mini-restaurant; or more radically, customers could be encouraged to select their choices from a set of colour monitors and keyboards networked into a central computer. The choices would be automatically queued (to the micro-second) into the kitchen. Obviously it would be a relatively simple matter to cost each possible combination, to print out an order docket for the till, even to monitor the stocks of certain raw materials. In an economic environment where successful service concepts can make so much money for relatively low capital costs, it is surprising that more thought has not gone into the idea of service design.

In the manufacturing field it has long been known that the secret of good quality is to design it in. Deming has pointed out that 94 per cent of faults are what he calls 'common cause' faults, i.e. they are built into the design of the system. He estimates that only 6 per cent of faults are finally traceable to worker error (Deming 1986). When American Express looked at their response time

CASE STUDY: COMPLAINTS HANDLING SYSTEM

The volume of someone's fury is not always proportional to the gravity of the situation. However, we endeavour nowadays to accept customer complaints and not to understate or fight them and try and figure out why that customer, person, area or section is 'making up problems'. Our Complaints Handling Process is not a 'finger pointing' exercise as proper recording of all complaints is a prerequisite to any formalised Complaints Handling system.

We analyse customer complaints for maximum information. In this context, we have recently been in a position to enhance the system with two new developments.

- We use available information technology selectively to help us through the necessary detective work in problem identification.
- Reports are reviewed by myself and then referred to the training supervisor in order to consider the relevant changes to the various processes that may be required. These are also copied to all supervisors for open discussion at our regular fortnightly meetings. Resultant changes to either process or procedures are then communicated by each supervisor to their own customer service teams.

Key features:

- Our definition — 'A legitimate complaint is one where, any of our customers, external or internal, record their dissatisfaction with any aspect of the departmental output or service levels.' All written complaints are formally logged each morning and the details recorded on a data base.
- Same day acknowledgments are issued where appropriate. Complaints are handled as a number one priority and immediate action is required.
- Complaints are expected to be finalized and the outcome communicated to the customer within ten days. The necessary corrective action is again recorded on the data base.
- The details are retained by the training supervisor.

Measurement:

- Monthly reports are produced and reviewed by the manager and training supervisor and discussed with all relevant supervisors and staff.
- Since this information became available to us at 1 March 1990, we have established that 90 per cent of all complaints are satisfactorily handled within ten days.

The total number of complaints recorded within the member administration department for the first nine months of 1991 shows a reduction of 22 per cent over the same period for the previous year. However, these cannot be ignored and using the Pareto analysis technique, we will be applying the quality improvement process to these items.

Source: VHI Member Administration Department

to cardholders' queries, they noticed that although the queries were allocatable to a small number of types, in each case an individual letter had to be drafted, typed, corrected and sent. When they composed 13 new form letters, and batched similar correspondence through the word processing department, response time was cut from 16 days to 10. Note that what had changed was the design of the service procedure. Like the man in Molière who was thrilled to be told he had been talking prose all his life, whether willed or not every service process already has a design. It may be a bad design, or at least subject to considerable improvement, but there is a design in the sense that there is a deliberate, planned, order of activity. Total Service Quality analyzes this design to see if it can be improved.

New ways of looking at service

1. The players: Normally company structures are displayed, and thought of, as a straightforward hierarchical structure. At the top is the board, then the chief executive, and below him the executive directors; reporting to them are various middle managers, junior managers, supervisors, shopfloor workers. This model certainly reflects the line of command. (It is no coincidence that a military phrase comes in here. The basic hierarchy model, like much other primitive management thinking, was derived from military models.)

For service companies, however, it is more fruitful to address the structural question differently. The different method inverts the pyramid. At the top layer (what was once the base of the pyramid) are the customers; just below them come the frontline service people, and below them the support units. At the tip of the inverted pyramid comes the management team, whose function is not to bark commands, but to support and steady the service providers.

This orientation changes the focus of the company from internal chain-of-command concepts derived from the army to outward looking customer servicing concepts. It is what scientists call a paradigm shift, comparable to that of the sixteenth century which changed the view of the solar system from being earth-centred to sun-centred. Paradigm shifts of this sort have impacts well outside their original arena. Darwin's theory of evolution for instance was originally intended simply to explain the relatively technical problem of how there came to be so many similar but different species of animals and plants; its impact on religion, on people's image of themselves and on science was profound.

The first change induced by turning the hierarchical pyramid on its head is in the self-image of management. Instead of managers perceiving their job to dictate, control and drive workers, they are seen as supplying the structures and materials that will enable them to do the crucial job. Instead of managers perceiving themselves as driving the workers like so many erratic and wilful bullocks down a lane, they become trainers, coaches and planners for a team of players.

Each level of activity is now perceived as supporting the frontline people in delivering to the customers that crucial moment of truth. As Karl Albrecht

1. STANDARD HIERARCHY

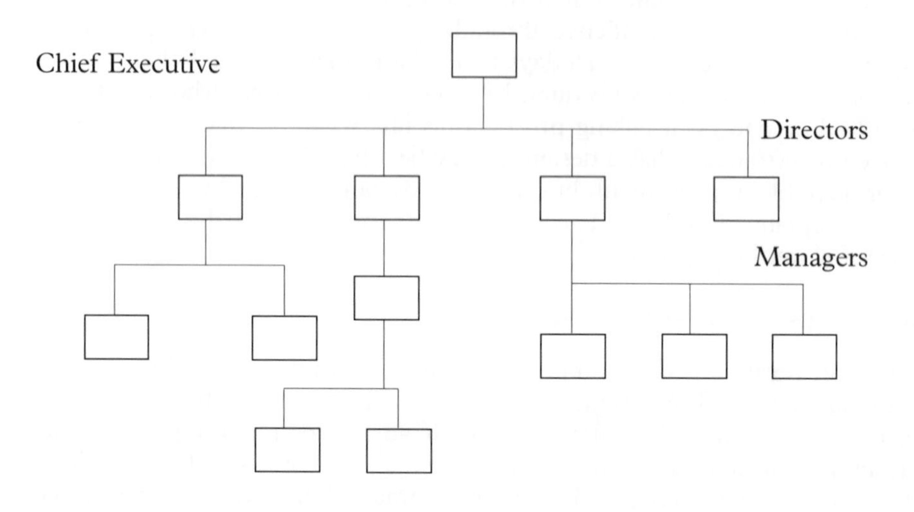

2. INVERTING THE PYRAMID

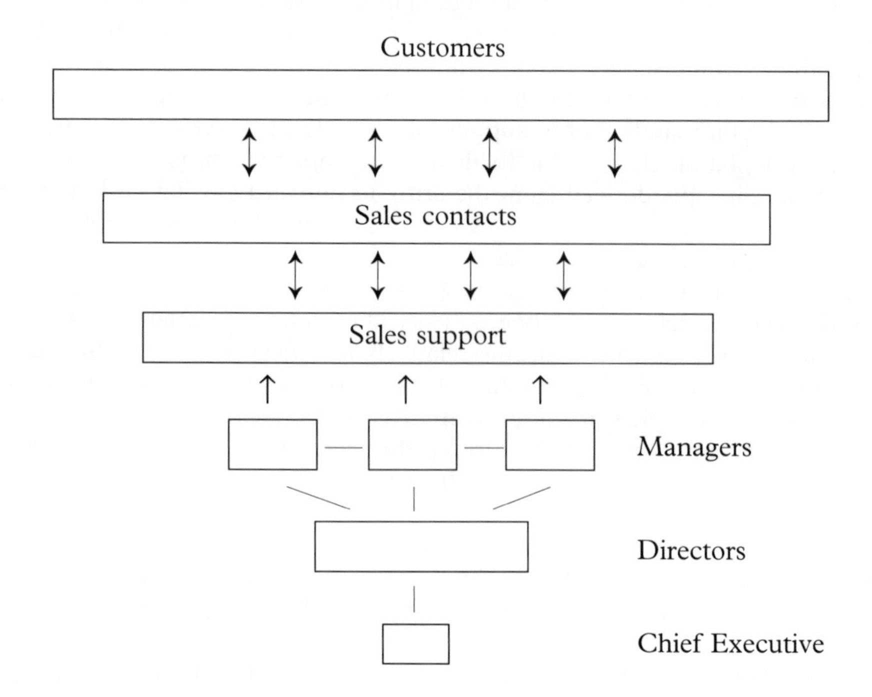

CASE STUDY: COMPANY ORGANISATION CHART

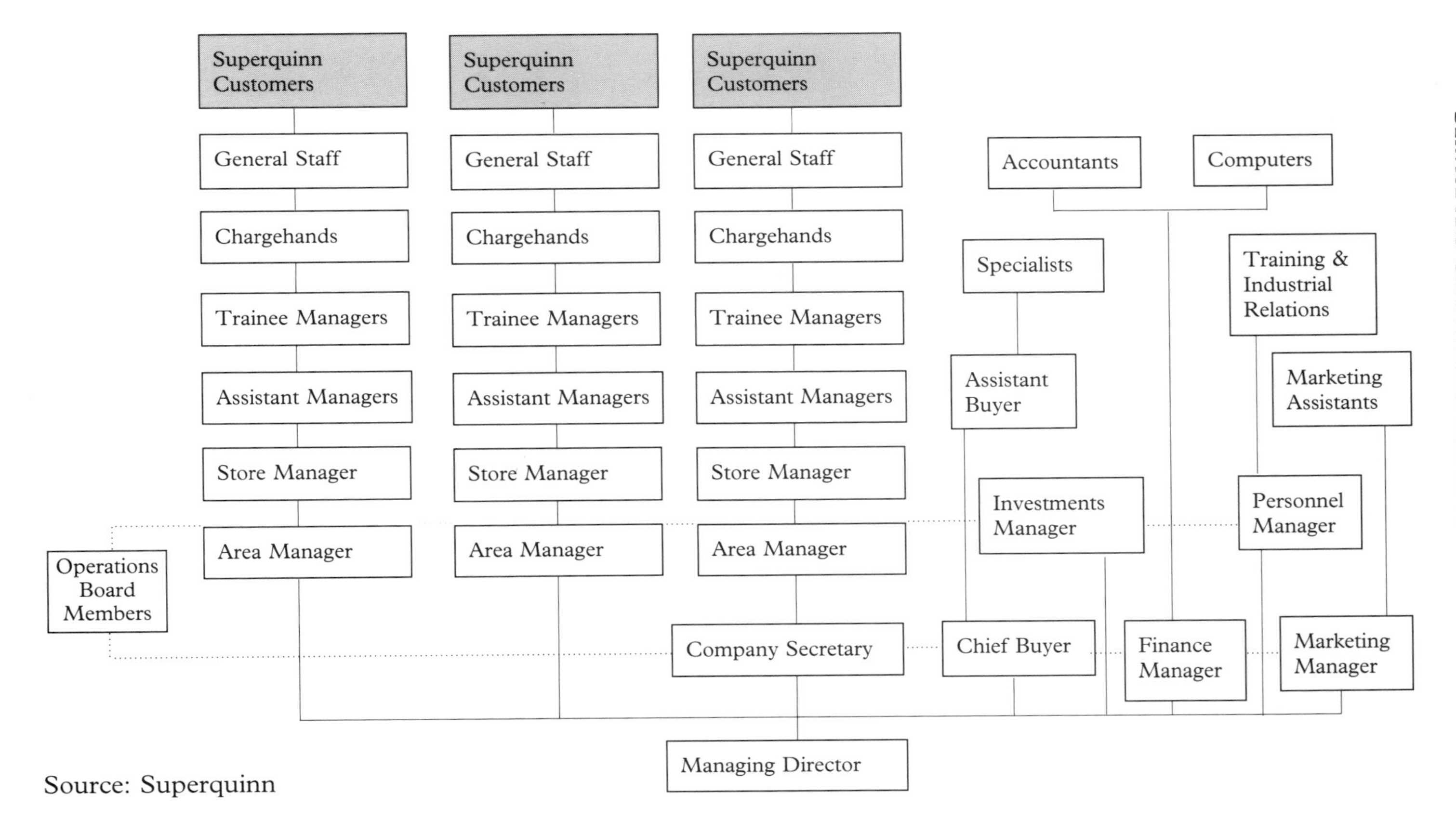

Source: Superquinn

put it 'the trouble with theories like the one presented above is they are easy and fun to think about, but they are the devil to bring into reality . . . managers will have to learn to live with a greater degree of employee autonomy, which implies a higher level of ambiguity for them as leaders. They will have to give up some of their reliance on generic rule-based solutions and be willing to tailor their leadership more to the variations in circumstances in the front line.' (Albrecht 1988)

2. The service blueprint: One way to start the analysis of the service is to draw up what Jane Kingman-Brundage calls a 'service blueprint' (Jane Kingman-Brundage 1991). The service blueprint is a combination of organization chart and flow chart. The key concept is that service is both a *process* and a *structure*; in other words there is both a manufacturing element and a positional element.

A service is a process because it is interactive between the customer and the service delivery person. This interaction distinguishes it from manufacturing. Yet it is also a structure: the kitchens and tables of a restaurant, the computers and branches of a bank, the bucket and swab of a window-cleaner are all familiar examples of structure, but there are other equally important ones, such as organizational structure, accounting structure and so on.

Obviously customers cannot interact with the service delivery people unless there is a structure (kitchens, bank branches, aeroplanes, etc.); nor can the structure do anything without the customers who create the process. Service design therefore combines the elements of process and structure. In a service blueprint, process is 'depicted from left to right on the horizontal axis as a series of actions (rectangles) plotted chronologically . . . a flow line marks the service path by connecting discrete actions. Service structure is depicted on the vertical axis as organizational strata, or structural layers.' (Kingman-Brundage 1991, 7)

These structural layers are normally divided into three layers, frontline employees, support staff and managers. As in the technique of turning the pyramid upside down, the organizational structure is also turned to focus on the customer-frontline staff interaction, with the support staff and managers seen as back-up forces. A series of lines divide one layer from another. These lines are the crucial points where service actions or support actions move from one mode to the other. They are the key activity areas for monitoring and control.

- the Line of Interaction: The frontline staff interface with the customers across the Line of Interaction; customer actions are above this line, staff are below. This is where the moments of truth occur. The arrows connecting the action boxes jump to and fro across the Line, denoting the interactive nature of the service path.

- the Line of Visibility: frontline staff with whom the customers deal are supported immediately by 'back-stage' personnel who are vital to the operation, but are below the customers' sight. These might include setting up aspects of the operation, following up after the customer has left, as well as providing inputs during the service. In a restaurant, the customers see the waiters, but

CUSTOMER SERVICE PLAN

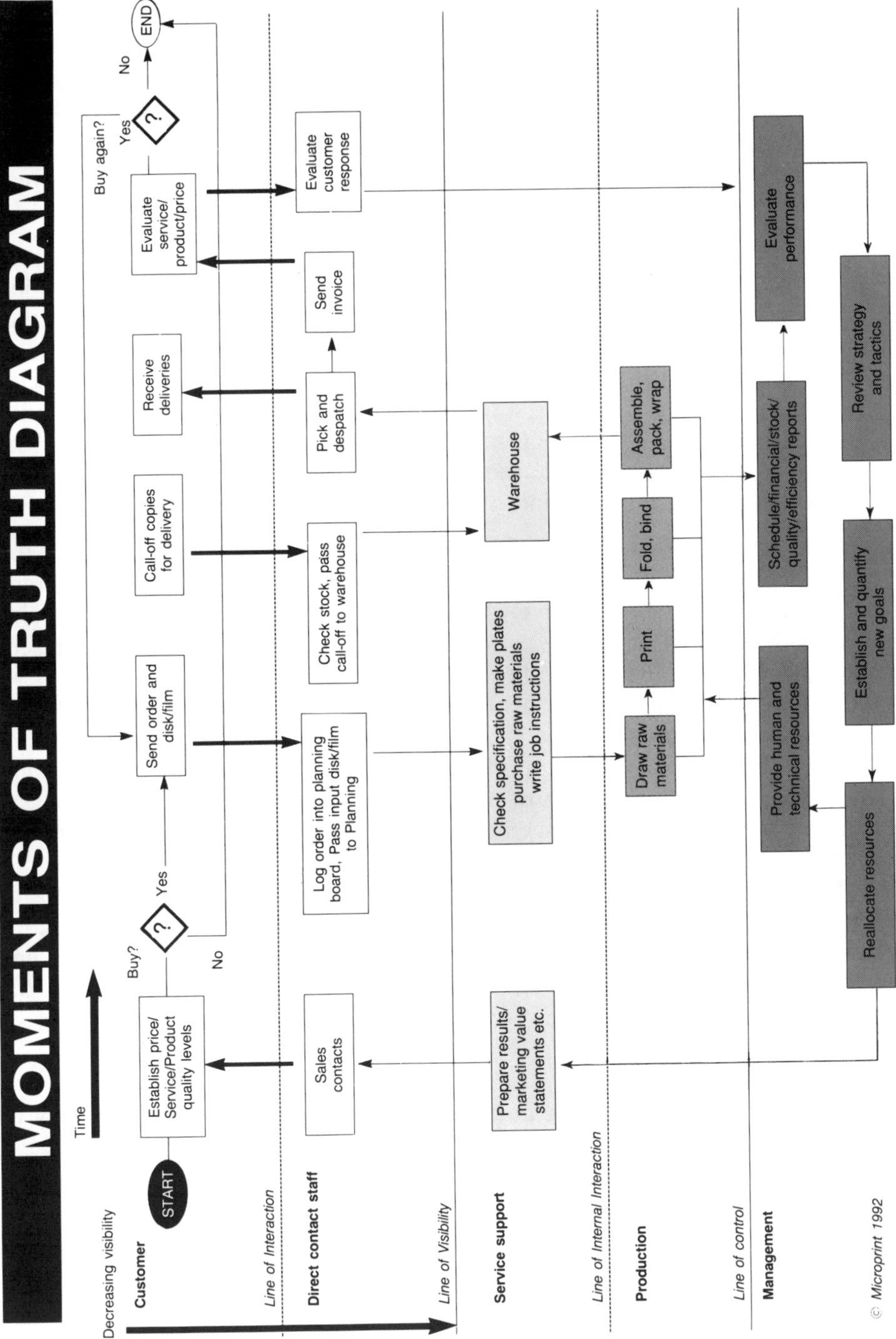

rarely the chefs. If you ring your bank manager to get a statement checked, he or she will handle the call, but will usually require information from some internal source. Not surprisingly, comments Jane Kingman-Brundage, the onstage actions often represent the tip of the business iceberg.

- the Line of Internal Interaction: Behind the immediate service providers, there are usually other internal operations, providing goods and materials to be used in providing the service. Their (internal) customers are the service providers. This picture of the service process enables all service employees to see exactly how they relate to the critical line of interaction.
- the Line of Implementation: This Line separates the 'doing' activities from the planning and resource allocation decisions. These boxes are not job descriptions, they are functions. It is possible, in fact it is desirable, that managers also operate in certain circumstances as frontliners at the line of interaction with the customers, as well as planning, coordinating and allocating resources. Bank managers, for instance, do this all the time. The service system blueprint should not imply that managers should lurk behind the line of implementation like the generals in the First World War in their chateaux fifteen miles behind the battle lines.

Kingman-Brundage explains how to read a service blueprint. 'Begin at the line of interaction. Who does what to initiate the service process? First trace the service process by following the service path from the consumer's point of view. Note any branching . . . existing or potential branching may present an opportunity to distinguish the service competitively.

Second, trace the contact person's path. Note the set-up and follow-up activities in the backstage area.

Third, trace the path of support functions, checking the points of initiation and termination and noting how and when hand-offs occur.

Finally, examine the management stratum, paying special attention to the management information reports.'

The service blueprint is a powerful tool in the development and analysis of the service delivery process. It is particularly powerful in pointing out the connections between the internal service suppliers and their customers, the frontline staff. In this model everything is centred on the Line of Interaction, the moment of truth line.

ACTION

- **identify the weights your company gives in practice and in theory to the five key factors in service quality identified.**
- **create a flow chart of the business from entry to exit.**
- **identify service characteristics.**
- **explore the various failure modes and their relative seriousness. (If in doubt, guess!)**
- **establish standards of acceptability.**

SUMMARY

This chapter identified five key factors in good service. These were, in order of importance: reliability, responsiveness, assurance, empathy and tangibles. The service delivery system contains both manufacturing and service elements: these should be analysed in the service flow chart. The key failure modes are identified, and a method of estimating the annual cost of each failure is presented. This enables management to pinpoint the most serious problems. The concept of service design is explored. What are the implications of changing from a boss/worker relationship to a coach player? The service blueprint, a new way of analysing the interface between service-giver and customer, is described.

CHAPTER 5

Employee Involvement

'Treat people as if they were what they ought to be, and you will help them become what they are capable of being.'
Johann W. von Goethe

Everything a service organization does is done through people. Despite this self-evident fact, in practice people are often prevented from delivering of their best. 'Personal quality', everyone agrees, is at the heart of the quality revolution but in practice the systems, documentation, slogans, command structures, anything and everything conspire to avoid facing the point. Compared to the effort put into cost reduction, the commitment to personal improvement is often meagre.

A US survey of workforces in the early 1980s found that less than a quarter of employees answered 'yes' when asked: Do you always do your best? Half of those asked said they did not put more effort into their jobs than was necessary to keep in employment; three-quarters admitted that they could be much more effective than they were usually (quoted in Møller 1988). This survey pinpointed a dramatic gap between what people actually did, and what they felt capable of. The problem is a deep one. It is not sufficient simply to utter a few well-meaning slogans. Attitudes to work, to motivation and to cooperation are mixed up with people's complicated attitudes to class and status and sometimes even gender. These are fundamental factors in a person's self-image, and are not easily changed. Most of the time structures and ideas set up in rather different circumstances are accepted as the comfortable and familiar ways of doing things.

Personal quality

Perhaps the first step is to recognize an important truth. In the absence of special pressures, a worker will behave at work as he or she does at home. A hard-working, considerate and honest person in daily life is likely to be the same at work. The personal ethical quality of the workforce is therefore not an irrelevance or a luxury in corporate terms: it is the basis of all other quality. This of course is especially significant in service industries, which demand so much of personal qualities.

Business is a cooperative exercise: the goal is to make more money and have more fun together than we could separately. Hard-nosed, old-fashioned employers however often fail to treat their subordinates in this way; too often work is a more or less muted form of exploitation. As the American writer Studs Terkel found in his many interviews with ordinary working people: 'for the many there is hardly concealed discontent . . . "I'm a machine", says the spot welder. "I'm caged", says the bank teller, and echoes the hotel clerk. "I'm a mule", says the steel worker. "A monkey can do what I do", says the receptionist. "I'm an object", says the high fashion model. Blue collar and white call upon the identical phrase: "I'm a robot."' (Terkel 1974)

Because these workers have no respect for the job they are doing, their self-esteem falls. Then their ability to provide the high level of service required in the modern economy falls also. With the loss of self-esteem comes a loss of commitment, a negative attitude, and sense of being a victim of life rather than a creator of opportunities. It is impossible for anyone suffering this loss of self-esteem to deliver high quality.

Product or service quality, says Claus Møller of TMI International, critically depends on the quality of the workers in the department; this in turn is dependent on the personal quality of each worker. Møller carries his analysis further by establishing two standards of behaviour to evaluate personal quality. These are:

- the AP level: the actual level of day-to-day performance.
- the IP level: a person's own ideal level of performance.

Satisfaction with personal quality will be achieved when the AP level approaches the IP level. To stimulate thought about the two levels, Møller poses various ordinary situations, and asks his readers to imagine their AP and IP levels of response. For instance:

- when you leave a shop, you realize that a cashier has given you change for a twenty pound note not a ten pound note.
- you slightly dent a (new) parked car.
- your child comes home delighted with a new toy, which you discover was acquired dishonestly.
- a mistake made by you was blamed on a colleague.
- a dinner guest, who has clearly had too much to drink, is planning to drive home.

- a harmless, but belittling, piece of information about a colleague comes to your attention.

In each of these cases there is a range of possible responses from the mean, self-centred, or dishonest to the idealistic and deeply committed. A person's ideal response is very often higher than the typical response. In analyzing the gap between AP and IP levels both for oneself and for others, the gap between the two levels is likely to appear unusually large or small in particular circumstances. This might be:

- when with certain people.
- when performing certain tasks.
- when experiencing certain types of reward/punishment.
- when exposed to specific influences.
- when in certain situations.

The challenge is always to bring the AP level up to the IP level. This should be done by establishing personal quality goals; by assessing the AP/IP level of recent actions (what the Church calls examination of conscience, in effect); by checking the effects of your actions on others; and by maintaining high personal values such as consideration, self-discipline, ethical behaviour and quality consciousness for oneself and others.

Certain guidelines can be distinguished to enable an AP level to be brought up to the IP level. For instance:

- determine personal quality goals.
- find out to what extent colleagues are satisfied with your efforts.
- treat the next person in the process as a valued client.
- avoid errors, and perform tasks in the best possible manner.
- make the best use of resources.
- be committed, take pride in your work (if you cannot, ask why).
- finish what you start.
- be ethical and maintain integrity and dignity.
- silence is equal to consent.

The division of labour

At the root of the problem of involvement is the division of labour. Early in the industrial revolution the economist Adam Smith perceived that the division of labour is the key to economic success. By allowing each person to specialize in a particular function, society as a whole benefits by having myriad specialists, each of whom is better at the task than any generalist could be. Naturally industry exploited this discovery. In his famous example of the making of nails (then done by hand), Smith pointed out that a common blacksmith without special training could scarcely make 200 nails a day; a practised smith could possibly make 800 or 1,000; but otherwise unskilled boys who did, and could do, nothing else could by specializing make as many as 2,300 in a day.

Later, the division of labour enabled specialized machines to be devised, it reduced the amount of time spent in passing from one kind of work to another, and by practice (the operation of what we now call the learning curve) greatly increased the skill of the operator (Smith 1776).

There is no doubt that the ability to specialize, the division of labour, has produced extraordinary benefits. It has also proved difficult to manage. Those boys capable of making two thousand or more nails a day at 18 had no other skills at 48. Since they were producing no more, they were worth no more to the factory owner, despite by then having a wife and children to support. Moreover the demand for nails was a function of the notoriously variable building demand curve. It was bound to wane from time to time, leaving the nailmaker with no wages and no other skills to fall back on. Even in good times, clanking out 2,300 nails a day hardly provided much job satisfaction. No wonder industrial bitterness arose. Workers showed their resentment of the conditions imposed by extreme interpretations of the division of labour theory by absenteeism, strikes and by bad-quality work.

Theories of motivation

To counteract these negative side effects of the division of labour, managers and consultants began to explore industrial psychology. Experiments in the Hawthorne factory in the 1920s and 1930s suggested that there was more to performance than job design and innate ability. As one modern textbook put it 'the results of their experiments emphasized that the worker is not a simple tool but rather a complex personality interacting in a group situation that is often difficult for managers to understand'. (Hellriegel, Slocum, Woodman 1989)

Three factors seemed to be at play in work performance. These were innate ability, training and motivation. Little could be done about innate ability, except by ensuring that the right people were recruited in the first place (the techniques of recruitment will be discussed in more detail in Chapter 6). Training was split into job design and straightforward training, and much effort was put into both. Indeed the famous (and now much abused) School of Scientific Management founded by Fred Taylor in the US spent much effort on details of job design. Production efficiency could be greatly enhanced by close observation of the individual worker, and by eliminating wasteful time and motion in his operation. In one famous case, Taylor devoted many pages of one of his books to explaining how a navvy should use a shovel.

Much theoretical work was also done in addressing the question of motivation. It was perceived that the motivational process begins with identifying a person's needs. These can be seen as deficiencies experienced at a particular

CASE STUDY: INDUCTION OF A NEW RECRUIT

APPENDIX XXIX: INDUCTION CHECKLIST

Name: Job Title:
Date of Commencement: Clock No:

Induction Checklist

Please tick the boxes if the information has been given/received.

☐ 1. Explanation of tasks, duties and responsibilities.
☐ 2. Confidentiality and conflicts of interest.
☐ 3. Clock card procedure.
☐ 4. General tour of premises.
☐ 5. Introduction to colleagues.
☐ 6. Location of staff facilities [canteen, cloakrooms, lockers, smoking areas].
☐ 7. Time-keeping and duration of official breaks.
☐ 8. Safety hazards.
☐ 9. Explanation of Company Safety Policy.
☐ 10. Fire/Emergency procedure.
☐ 11. Location of First-Aid stations.
☐ 12. Identity of Safety Committee members/Safety Representative and trained First-Aid personnel.
☐ 13. P45 submitted to Quality & Systems Administration Manager.
☐ 14. Payment of wages.
☐ 15. Jefferson Smurfit Group Ireland, Employee Handbook
☐ 16. Commitment to Quality and The Educational Company of Ireland Quality Manual.

Your standard contract of employment [written Notice of Terms of Employment as required by the Minimum Notice and Terms of Employment Act 1973] will be issued within 7 days. If you have not received your contract within this period, please bring the matter to the attention of your immediate superior.

Signature of Employee: ________ Signature of Superior: ________

This form must be completed on the first day of employment and sent to the Chief Executive.

Source: Educational Company of Ireland

time. They might be psychological (such as the need for self-esteem), or sociological (such as the need for friendship) or physiological (such as the need for air, food, water) (Hellriegel, Slocum, Woodman 1989, 141–73). Abraham Maslow ranged these needs into a hierarchy ranging from the lowest physiological needs to the highest self-actualization ones.

He identified five categories of need. These are:

Physiological needs: The basic physical needs for air, food, water, shelter.

Security needs: Once these are satisfied, the worker seeks to ensure that they will continue; therefore he seeks safety, stability, absence of pain, threats or illness etc.

Affiliation needs: The needs for love, friendship, and a feeling of belonging. People are basically social, and this need is evidenced in the strength of factory social clubs, etc.

Esteem needs: Everyone likes to be valued or esteemed by people they respect, by their colleagues and others. The desire to meet such needs has led to the development, particularly in the US, of various devices to promote employees' pride in their work.

Self-actualization needs: The need to express creativity.

Maslow's view was that, broadly speaking, a satisfied need does not continue to motivate. A thirsty person will do much for a drink; once the drink has been taken, however, that need is satisfied and no longer operates as a motivator. The possibility of motivating this man does not cease; the motivation 'trigger' simply moves to a higher level. In general, lower level needs must be satisfied first before higher needs are likely to be considered.

Other writers have taken the hierarchy of needs concept and refined it in various ways. Frederick Herzberg of the University of Pittsburgh, for instance, regards the motivational effects of needs as falling into two broad categories, which he called hygiene factors and motivating factors. Hygiene factors such as supervision styles and basic working conditions of work do not motivate; they simply provide a suitable base for motivated work. Motivating factors, which broadly take up the top three elements of Maslow's hierarchy, are the ones that persuade people to work well. Salary is a hygiene factor, but also a low-level motivator.

Herzberg and his colleagues found that when people felt positive about their jobs (i.e. when the motivators are in good operation), they put more care, imagination and craftsmanship into their work; when they feel negative, they do no more than comply with the minimal requirements.

These two theories of motivation are the best-known examples of the general category called Content Theories. They concentrate on identifying how specific factors motivate people. The manager is then left to activate these as best he or she may.

Another set of theories, called Process Theories, takes a different view of the problem of motivation. They are concerned with how workers make choices. They ask the question: 'What makes outcome A more desirable than outcome B?' Process Theories assume that people make rational choices based (consciously or unconsciously) on the possible outcomes of a course of action and the relative probabilities, based on past experience.

A specific type of Process Theory suggests that the strongest factor in such decisions is not the likely outcomes, but how the individual feels about whether he or she is being fairly treated. Resentment against unfair treatment relative to others pushes the worker into negative attitudes. Feelings of being unfairly treated were found by Herzberg to be among the most frequently reported source of job dissatisfaction.

All of these views are expressed from the management side. Managers (implicitly *us*) have a responsibility to the shareholders to make sure that each of the potential of these assets is maximized. In particular, managers have a responsibility to ensure that the workers (implicitly *them*) jump out of the trench-es when ordered to and charge the enemy as vigorously as possible, impelled by little more than *esprit de corps* and an automatic reaction to the word of command.

Should the monkeys run the zoo?

The problem with this attractively simple approach is that it does not reflect what is happening in the rest of western society. The Japanese quality revolution gains much of its strength from exploiting the social and political structures of Japanese society. So much so indeed, that some commentators have wondered if the quality levels achieved routinely by Japanese companies are possible at all in our very different society. The failure of the Quality Circle movement in the early 1980s, despite many brave attempts, simply reinforced that suggestion. It is said that western society is too individualistic to cooperate in this way.

Certainly the individualistic approach is deeply ingrained in western society. The only acceptable way to run a country, for us at least, is by some form of democracy. Once out of the factory or work-place, every person's opinion is eagerly canvassed and has equal weight, if only at election time. Heavily planned, centralized economies simply do not work as well as free market ones. It has taken some time for this view to prevail. For much of this century people have hoped that although political dictatorship does not work, economic dictatorship might at last spread the fruits of the nation's work evenly to its citizens. In country after country, more or less rigorous forms of centralized economic planning and ownership by the state of the commanding heights of the economy was initiated.

In public life political and economic freedom works best for us. Only in business life does the authoritarian style reminiscent of the landlord/peasant relationship of history prevail. Of course modern work-places are not as grim as they used to be. Wages are higher, hours are shorter, operations are less hazardous; offices are brighter, warmer and quieter. But the basic relationship between the employed and the employer remains: a kind of voluntary indenture, in which the worker agrees to submit to supervision and control in exchange for an agreed wage. Outside the business workers can spend their money as they like, vote as they like, express their opinions. Inside they are (generally) 'not paid to think', as the army used to say.

The Chairman of Eastern Airlines in 1986 expressed a common (though not usually so bluntly expressed) senior management view about worker participation, when he said, 'I'm not going to have the monkeys running the zoo.' An economy works best if it is not dominated by the top: a business apparently works best if it is. This culture of authoritarianism is so deeply embedded in our organizational thinking that it is difficult to imagine how things might be different.

This contrast between the culture of business and the preferred modes of public life was very apparent to serious thinkers at the beginning of the Industrial Revolution. In the United States, Thomas Jefferson feared that the dependency of mind induced by the factory would undermine the nascent republic. 'Capitalism was incompatible with republicanism', many Americans believed. 'Dependence on wages robbed a man of his independence . . . the boss was like a slave owner; he determined the hours of toil, the pace of work, the division of labour, the level of wages; he could hire and fire at will.' (McPherson 1990). How could a man who was dependent on a boss for his daily living, his access to the machinery by which he fed and clothed his wife and children, behave self-reliantly in national affairs?

We have seen that the Japanese success with Quality Circles and other techniques has been partly derived from their social and political culture. In the West however we have decided that workers can be democratic and free outside of the factory, but once inside must behave as the company's 'greatest asset', and be controlled accordingly. In doing so we have established a split in interests between employed and employer that militates against the ultimate objective of the organization, which is to make money by providing a service to its customers.

The deep questions of corporate culture

Modern thinking on organizational style has concentrated on distinguishing between autocratic and aggressive styles of management, often called Theory X management, and the more person-centred style called Theory Y. Theory X is based on the military ideal exemplified by the centurion in the New Testament: 'I say to this Go, and he goeth, and to another, Come and he cometh, and to my servant, Do this, and he doeth it.' (Matt. 8.9.) Theory Y attends to the human needs and motivations of the individual. Most writers believe that some derivative of Theory Y is best adapted to make the most of modern workforces.

One writer, Rensis Likert, for instance, has identified four systems of management, ranging from what he sees as the least to most productive style:

- exploitative authoritative.
- benevolent authoritative.
- consultative.
- participative.

The more the organization moves from a command style to a participative style, the more efficient it is likely to be.

The quality of service is peculiarly dependent on personal relations between the service giver and the customer. The self-image of the service giver is critical. In environments where the bosses have separate toilet facilities from the ordinary workers, where they use different tables in the canteen (or even in extreme cases, the partners/executives' dining rooms), where they work different hours and wear different clothes (suits as opposed to uniforms or overalls), it is difficult to imagine that the uniformity of purpose urged in Chapter 3 can be easily achieved.

At the very beginning of the quest for Total Service Quality, aspects of the firm's culture will begin to be exposed. In the early meetings, most people will be enthusiastic, excited, ready to take part. The total quality process makes use of all the energy at present locked inside the system. If people are not asked or encouraged to give their opinions, by and large they will not. In the absence of a proper process, what opinions they do give are often based on poor information and so fail to take important facts into account. Such is the state of our service industries, that for many people even to be asked their opinion is an exciting change. To be brought into a process that institutionalizes this can reawaken all sorts of dramatic contributions.

Most people will find this exciting. A few will ask: 'What's in it for me?' The benefits to the bosses are clear; the benefits to the ordinary Joe may not be. At this stage you have struck the first challenge to your sincerity as a manager. The introductory process no doubt mentioned change and acceptance of change as fundamental to the Total Quality process. Now, early on in the process, one of the most fundamental changes has been asked for. Whether the management will deliver on this is a touchstone for how seriously the company actually takes the proposed new approach. If the Total Service Quality programme is ultimately seen as another nifty way of getting more out of the workers, it will quickly fail.

There are three fundamental problems in the background culture of any company that will be challenged during the Total Service Quality process. These are:

- how the company shares its income.
- the division of labour.
- how dissidents are handled.

How the company handles these questions dictates much else in the company's culture and ultimately in its ability to serve its customers.

Who shall have this?

The first question is that of sharing the spoils. A company is fundamentally a group that has come together to increase the incomes of all by exploiting the principles of the division of labour. The question immediately arises as to how the group income shall be divided. Clearly the simplest solution would be that everyone should be paid exactly the same. Hardly anyone will suggest this as practical: but if not, how are the relative contributions of the teaboy, the marketing director and the skilled worker to be evaluated?

Since few of us have time to re-invent the wheel, it is lucky that the question is usually decided by convention: those at the top of the management hierarchy, particularly if they are also the owners of the company, get the most, and others get less depending on their hierarchical position. Indeed it is axiomatic that a subordinate should normally be paid less than his or her boss. There is little evidence that this represents real contribution, so many companies have developed incentive schemes that relate payment to results. Sales people working on commission in particular are frequently paid more than the top managers in companies.

In a democratically oriented company, it is clear that contribution rather than hierarchical position should be the determinant of pay. Tom Peters suggests that as much as 25 per cent of base pay should be related to performance, for all workers in the company, paid at least monthly. (This incentive-based scheme should be, in his view, attached to significant security of employment.) Exactly how this should be calculated is another matter. Clearly the measures should be as transparent as possible, and visibly in the control of the particular workers or teams. (Peters 1987, 333–42)

The division of labour

The division of labour is the fundamental factor in the efficiency of the company. Workers combine and specialize in order to make more money together than they could apart. Historically, the ability to harness the efficiencies provided by the division of labour has enabled great empires to flourish, and corporations to prosper. But the division of labour has to be deliberately organized. Typically, not much thought is given to this question. Certain jobs are clearly for graduates, others for women, for craftsmen and so on. Workers are trained to fill a particular niche, and are expected more or less to go on filling that niche for ever. Tradition in the company and in the industry normally dictates who goes where.

The irreconcilable minority

The third question that has vexed countries and managements alike is that of the irreconcilable minorities. Just as virtually every country in the world has its own ethnic or national minority (usually with its equivalent of Kerryman jokes as well), every company will have an awkward squad. It is a measure of corporate culture how these dissidents are treated. In organizations that put a high value on conformity, dignity and status, dissent tends to be silent. This is very much the Japanese way; as their proverb has it: 'the nail that sticks up gets hammered'. In order to counteract the negative sides of this, the Japanese have put very high value on such techniques as suggestion schemes, brainstorming groups and mental mapping techniques.

The treatment of the awkward squad will depend very much on how authoritarian the company's culture is. Certain companies tend to favour the conventional dresser, the aggressive masculinity, 'toughness', in opposition to subjective, feminine, creative values. In these companies a submissive, uncritical attitude to the moral authority of the group is important; there is also a tendency to condemn outsiders. In these companies, as in armies, minority behaviours are neither encouraged nor tolerated. N.F. Dixon's study *On the Psychology of Military Incompetence* has many lessons for managers in this and other contexts. (Dixon 1976, 258)

Introducing change

Change is painful and difficult for many people. It requires a move from established and comfortable psychological attitudes to new and challenging ones. The path has therefore to be well planned and well laid.

In particular, managers must:

- listen to what people say, and to what they mean.
- recognize individuality and psychological vulnerability.
- motivate the unfreezing of existing attitudes.
- listen.
- identify and commit to new attitudes.
- treat employees as adults by involving them.
- listen.
- be flexible — there are more ways than one of killing a cat.

With the best will in the world, it is unlikely that a Total Service Quality plan will go through your company without opposition. People will resist change. There are some good reasons, and some not so good. Most of the reasons for resistance to change, however, boil down either to fear or to reluctance to be disturbed. For instance, workers (and this applies to all staffers, from MD to warehouse) may resist change because of fears, such as:

- fear of seeming incompetent, perhaps because of lack of required new skills.
- fear of loss of control — it has been well said that 'people do not hate change, they just hate being changed'.
- fear of uncertainty, especially surprises.

However, the reaction may be one based on a reluctance to change, perhaps because the job is no longer exciting, but just part of daily life. These reluctances might include:

- reluctance to change established habits.
- reluctance to face new challenges.
- reluctance to risk knock-on changes, to establish precedents.
- reluctance to take on extra work.

Whatever the reason for resistance, the key weapon against it is commitment of the senior people in the business to change. If the top managers are totally committed, then gradually even the most reluctant and fearful workers will follow. It is important that while expressing this commitment you do not attack the past; what happened in the past after all was created by those same workers you now need to change for the future.

The second rule is to abolish surprises by extraordinary amounts of talking and listening. The rule must be communicate, communicate, communicate; paradoxically perhaps, this can only be done by real listening.

Communication is a two-way process; it is easy for those in authority to forget that. A regal tour of the office or kitchen, perhaps with a kind word here and there, is not communicating, because there is no listening on your part. Listening in particular means being willing to ask silly questions. Senior managers often feel that they lose face by not knowing aspects of their workers' jobs; the myth that knowledge is the special preserve of management dies hard. The employee involvement aspect of the Total Service Quality process is based on the idea that everyone in the company has a special knowledge of their area. The plan is to unlock the power and insight that these special knowledges hold. Therefore listeners should not be afraid to break through jargon, to ask for things taken for granted to be explained, to ask why.

Finally, real listening is attending and reacting to what is being said, not simply waiting for your turn to speak.

ACTION

- **explore how the three key attitudes in corporate culture are addressed in your company.**
- **would the middle and junior members of the staff feel the same way about these points as you do?**
- **what are the forces for resistance to change in your business?**
- **how many hours a week do senior managers spend listening to front-line staff?**
- **what do your junior staff know about their jobs that you do not?**

SUMMARY

The achievement of employee involvement has to start with the very basic culture of the company. Anything else will be quickly perceived as window dressing. However, a culture that does address these problems will be on the high road to achieving that unity between organizational ideas and societal ideas that the Japanese have exploited so effectively. The contradiction between the two sets of background assumptions is a cause of alienation; if it can be removed, energy will flow.

CHAPTER 6

Recruiting and Training Staff

'There are no limits to the ability to contribute on the part of a properly selected, well-trained, appropriately supported and above all committed person.'
Tom Peters *Thriving on Chaos*

In service industries, people are the key to success. Consequently, they should be planned for more carefully than any other facet of the service design. Unfortunately, human relations being what they are, key appointments are too often decided on the golf course, with no checking of references, or even perhaps a good idea of what old Tom's nephew is actually going to do!

The ISO Guidelines for Service Industries gives a clear message on the requirements for personnel policy and management to achieve the firm's quality objectives. Management should:

- 'select personnel on the basis of capability to satisfy defined job specifications.
- provide a work environment that fosters excellence and a secure work relationship.
- realize the potential of every member of the organization by consistent, creative work methods and opportunities for greater involvement.
- ensure that the tasks to be performed and the objectives to be achieved are understood, including how they affect quality.
- see that all personnel feel that they have an involvement and influence on the quality of service provided to customers.
- encourage contributions which enhance quality by giving clear recognition and reward for achievement.

- periodically assess the factors which motivate personnel to provide quality of service.
- implement career planning and development of personnel.
- establish planned actions for updating the skills of personnel.' (ISO 9004–2: 1991(E) *Quality management and quality system elements Part 2: Guidelines for services* 5.3.2.1)

The objective of a recruitment and training programme is to build an effective service quality team. To do this, the management team must:

- design quality into each job.
- write job specifications using established service standards.
- screen applicants carefully for service quality strengths and abilities.
- use effective training to maximise the development of service skills and aptitudes.
- practise the principles of leadership that promote winning teams.
- nurture an organisational environment that supports customer driven service standards.

Planning staff needs

In any environment close attention should be paid to the personal quality of all staff. The formal requirements of long-term personnel planning require other areas to be attended to. Specifically, the planner must investigate:

1. The establishment: The first task is to assess the present personnel resources. Do we have enough of the right kind of people to do the job we want to do? If staff are working large amounts of overtime, for instance, perhaps the firm is understaffed. Long working hours make it difficult to sustain high quality levels. This manpower audit will also explore such issues as the age structure of employees, the training levels and the trends in absenteeism and sickness to identify the current position.

Even if the staff/demand equation is balanced for now, the balance will change. Is the firm shrinking or growing? If so, how quickly? Plans for expansion or for branching out into new areas of activity may require new skills, new types of background; some of these may be catered for by training present staff. Others will have to be bought in.

This leads the planner into an exploration of the outside market. In some areas skill shortages can provide serious problems; in others an established pool of skills can attract new firms. The computer industry has grown in Ireland at least partly because of two sets of skills: the workforce's learnt experience with other computer companies in the recent past, and their ability to speak English.

2. Planning adjustments: A perfect balance between supply and demand is unlikely to be achieved or maintained. The task is first to identify why imbalances occur, then to devise appropriate corrective actions. For instance, a high staff turnover in a particular area may be caused by pay and conditions slipping out of step with what is available elsewhere. A high instance of union activity in another area may spotlight areas where management has failed to identify the workers' and the firm's mutual interests. Absenteeism is a well-known indicator of lack of motivation and also of work stress. It is for the planner to discover whether this is caused by the way the work is distributed, by method of supervision or perhaps difficulties with particular equipment.

3. Implementing solutions: The exploration of staff resources and the causes of imbalances and other problems may have made clear the necessity for policy change in areas such as:

- recruitment planning.
- training planning and programmes.
- industrial relations.
- management and employee development.
- pay and conditions.
- organizational issues.
- management attitudes.

These policy changes may involve changes in the organization's approach to job design. The service designer may have started with the *job engineering* assumption that the purpose of job design is to maximize efficiency, using time and motion information. This however can lead to employees becoming tied to machines or particular posts. Boredom, loss of motivation and then loss of quality quickly follow.

This may lead to management seeking to establish *job enlargement*, whereby more tasks are added to the job to create variety, or *job rotation*, which concentrates on moving skilled workers from job to job to ensure continued motivation. A broader view, based on Herzberg's theories of motivating and hygiene factors is called *job enrichment*. Four concepts have evolved as central to the job enrichment strategy; these concepts are important in the development of a quality system. They are:

- client relationships: workers should be put in touch with the users of their products; too often they wind up working for their bosses, not their clients.

- scheduling of own work: the individual or group of workers knows very well (better than management) how long things take, and the best order of activity; it is too often only a lurking distrust that obliges managers to retain the scheduling function.

- ownership of product: workers should be identified as closely as possible with the final outcome of the firm's effort.

- direct feedback: workers should be told regularly how they have performed. They should be encouraged to check their own work, to take responsibility for corrective actions and for designing better operating systems.

Another approach to motivation is to establish a *sociotechnical* system. In this, the group of workers are themselves responsible for the job design, and for balancing the social and technical aspects of the job.

4. Monitoring the solutions: Shifts in these policies go to the heart of the management of the company. Sometimes managers are following policies which they have barely articulated even to themselves. As a result they are difficult to change. It may take years for a change in recruitment policy to show any effects, and even then the results may be ambiguous.

However as with any quality policy, the effects of changes must be identified and analyzed as thoroughly as possible. Perhaps this can best be done by identifying at the introduction of the new policy:

- what the expected effects are.
- what yardsticks are to be used to monitor the results.
- whether cost savings are expected.
- what systems are in place to enable this information to influence future decisions.

Recruitment

The planning of staff resources described above gives the management an idea of its recruitment and training requirements. The first stage is recruitment. This starts with a job description, typically based on a formula which breaks the job into five dimensions:

- purpose.
- duties.
- methods.
- standards.
- relations.

These dimensions answer the significant questions about the job.

1. Purpose: What is the purpose of the job?

- what is the ultimate service required?
- what is the desired end-result of the service?

2. Duties: What does the person actually do in this job?

- what are the most important duties?
- what are the secondary duties?
- how often are the duties performed?
- what is the nature and scope of decision-making?

CASE STUDY: TRAINING IN THE QUALITY MANUAL

Training

16.1 *Policy*

16.1.1 It is the Educational Company of Ireland's policy to provide training which ensures that all employees possess the knowledge and skills required to perform their functions competently.

16.1.2 Company Initiative: Training is given to individual employees to tackle new challenges identified by the company.

16.1.3 Employee Initiative: Individual applications for special training are fully considered.

16.1.4 The company recognises the importance of career development and provides financial assistance for further education and training which mutually benefits the individual and the company.

16.2 *Responsibility*

The Chief Executive has overall responsibility for company training.

16.3 *Induction*

16.3.1 There is an Induction Programme for every new employee on their first day.

16.3.2 The new employee's immediate superior co-ordinates the Induction Programme.

16.3.3 An Induction Checklist is completed and signed by the new employee and his/her immediate superior (see Appendix XXIX).

16.4 *Appraisal Reports*

16.4.1 An Appraisal Report System operates on an annual basis for the following personnel:

- The Management Team
- Department Heads
- Sales Representatives

16.4.2 The Appraisal Report System is useful in the identification of training needs.

16.5 *Special Training Requests*

Individual requests for special training can be made to the Chief Executive at any time.

16.6 *Training Plan*

16.6.1 The broad training rquirements of the company which necessitate external training are identified annually in a Training Plan.

16.6.2 The Chief Executive is responsible for the Training Plan which is prepared in consultation with the Management Team and Department Heads.

16.7 *In-house Training*

16.7.1 The in-house training requirements of each department are identified on an ongoing basis by the Department Heads.

16.7.2 Training Programmes which require financing must be approved by the Chief Executive.

16.8 *Records*

The Chief Executive maintains a record of Training Courses and Programmes completed by company employees.

Document reviewed and approved:

Signature:	Date:	Revision
Signature:	Date:	Page of

Source: Educational Company of Ireland

3. **Methods:** How does the person actually perform this job?

- what methods, practical skills and/or technologies apply to this job?
- what inter-personal skills are required?
- what are the general working conditions (i.e. place, hours, hazards, advantages, colleagues)?

4. **Standards:** What are the standards of excellence required in this job?

- what procedural standards apply to this job?
- what personal service standards apply to this job?
- how is the quality of the job measured?

5. **Relations:** How does this job relate to others?

- what internal and external contacts are involved?
- what are the reporting relationships?
- what are the rewards?

(Martin 1991, 41)

The result of this process is the production of a list of key performance factors which the applicant will be expected to deliver, and against which long-term performance can be measured. A typical list of performance factors might include:

- task and target achievement/reliability.
- flexibility/responsiveness to challenge.
- assurance — do internal and external customers trust this person?
- market knowledge/empathy with customers.
- professional expertise.
- industry/product knowledge.
- communication/selling/customer relations skills.
- planning.
- staff management.
- administrative and financial skills.
- technical skill and knowledge.
- controlling and developing work quality.
- problem solving/continuous improvement.

No one can be expected to score highly on all these factors: brilliant sales people, for instance, are often weak on administrative skills. Equally, good administrators tend to be weak on flexibility and empathy. Different jobs will require different values.

The next stage in the process is to describe the person who could fill the requirements outlined during the job description process. The ideal person will combine a particular set of skills, sets of knowledge and attributes. Thus an advertising agency would have entirely different skill, knowledge or attribute requirements for a copywriter, an account executive or a media buyer. The factor-by-factor description of the job forms the basis of this. In working out the job description, part of the process is to establish which performance factors are essential and which are attractive but not vital. A successful candidate must score high on the essential factors to stand a chance. If the essential performance factors have been identified in advance, there is less temptation to appoint an unsuitable candidate (i.e. one lacking credible performance in one of the essential factors) simply because he or she is the best of the bunch. This is nearly always a bad idea. Quite soon a picture of the ideal person evolves, and this can be summarized under a set of characteristics, as follows:

- appearance and age.
- education.
- skill and experience.
- intelligence.
- health.
- interests.
- motivation.
- personality.

Having fixed on a job description, and identified the ideal mix of skill, sets of knowledge and attributes to fill the job, the next stage is to find such a person. In most cases it is ideal to recruit internally; indeed many companies make this a rule, believing that it is a crucial part of management's function to plan and develop workers to fill posts as the firm grows and develops. In these circumstances, internal advertisement of the post will come first.

Some consultants suggest that the job particulars should be circulated to senior management in advance to identify the possibilities of internal promotion. A development of internal recruitment is to request suggestions from current employees of suitable people to fill jobs. Once internal sources fail, the company has to resort to advertising or contacting specific training grounds such as universities, employment agencies and headhunters.

Many companies are still quite casual when it comes to evaluating candidates, despite the present climate of employment law, in which the offer of a job has serious implications. It behoves employers therefore to take the function extremely seriously. In practice, references are frequently not followed up, qualifications are not checked, track records are not identified and confirmed. Typically in a written CV, dates may be fudged to cover over cracks between periods of employment, memberships or qualifications 'written up' to seem more valuable than they are, and achievements aggrandized. Despite the frequency of such claims, in practice very few people have a real input into 'corporate strategy' or 'profit accountability'. As a result of this casual approach to the CV, from both sides, people are appointed to posts they should not get, and others are tempted to mislead when presenting CVs.

The interview itself, the typical selection procedure, is frequently handled in an equally casual way by untrained managers, who frequently do not even have a checklist or an agreed set of criteria. Activity tests, group presentation exercises, practical information tests, psychological tests etc. are occasionally used, but they are time consuming and the results frequently difficult to interpret.

One test that is not as well known as it should be is the 'In-tray Test' for managers. In this candidates are given a typical in-tray of materials (letters, memos, messages) with the brief to clear the tray by taking appropriate decisions. Two or three hours are usually necessary for this task. The in-tray contains sufficient information about the company and the role to be played. Problems include operational matters on which a decision has to be made, matters to be delegated, irrelevant matters, unnecessary information, current concerns. After the candidate has dealt with the problems, the interviewer should spend time with the candidate discussing how the problems were addressed, and the rationale behind the decisions made.

This test is one whose results should be acceptable to candidates, since it attempts to replicate real conditions. Other tests, such as those in the wilder fields of psychological testing, lie detection, graphology and astrology, generally produce less valuable results than the historic method of employing old Tom's nephew just because of the relationship.

For various human reasons selection and recruitment is usually given less practical and detailed attention than the purchase of new machinery. An investment of £50,000 in hardware would be explored and analyzed and examined from all angles; on the other hand, quite junior staff are delegated to recruit new employees earning say £12,000 a year. The hardware specialist goes on courses, and visits sites to see the new machinery in action. The recruiter is frequently given little more than an empty office and a pad of blank paper. Yet since these employees are very likely to stay for up to four years or more, the investment is the same. Indeed the damage that can be done by an inappropriate recruit is potentially much greater than the worst a machine can do. In a Total Service Quality organization, recruitment and candidate selection should be treated like any other quality procedure:

- objectives should be set.
- plans laid out in detail.
- controls identified, established and monitored.
- results recorded.
- records analyzed.
- analysis fed back into the new plans.

Training

Outside business life, we all know that the player who trains hard and practises hard will always beat the similarly talented one who does not. Keen golfers, yachtsmen and card-players spend a fortune on books and videos and training sessions to improve their game. Yet too often in the much more difficult and sensitive world of business activity, training is regraded as a distraction from the real world.

Real men do not eat quiche, and they do not go on business training courses either. Employees who meticulously service and maintain their cars sometimes resent spending money improving the skills of the workforce. Sending employees on courses, so the theory goes, is simply improving their qualifications so that they can find a better job elsewhere. 'After all,' they say, 'I'm running a business, not a university. If I employ an accountant, solicitor, waiter, computer programmer, I employ them to get on with the job, not to be improving themselves at my expense. I did not get where I am today by going on training courses.' This head-in-the-sand attitude accounts for the very low training budgets in most Irish companies. In Japan firms plan to spend up to ten days a year per employee on training. In most Irish firms the figure is less than one day a year.

Although management is admitted to be the most complex aspect of running a company, all skilled craftsmen undergo a longer and more arduous training than most managers. Most Irish managers are incapable of handling the level of statistical analysis recommended for shopfloor quality circles by Kaoru Ishikawa in his book *Guide to Quality Control.* Although queues and queuing are a vital practical part of service quality design, few service managers have a working knowledge of the well-established mathematical theory of queues. In management discipline after management discipline, from personnel selection, statistics, operational research, law, cost analysis and accounting, to market research, new product design, failure mode and criticality analysis, organizational design and computer applications such as database management, far too few Irish managers have had extended formal training.

If this is true of managers, it is also true of workers and supervisors. Men and women are shoved into functions with an absolute minimum of information and expected to 'pick it up as they go along'. At best there might be a bit of more or less informal on-the-job instruction. Neither of these favourite training techniques amount to more than a day or two a year: a far cry from the ten days achieved in major Japanese companies.

At bottom, the lack of belief in training derives from a profoundly static, short-sighted view of the business world. Despite daily evidence to the contrary, too many Irish managers (and the same proportion of English and American managers) appear to believe that the world of service and business is not changing constantly.

In sport, technology and politics new records and systems evolve all the time. In business, an accountant is an accountant: a cook a cook: a waitress a waitress: a manager a manager. They get their training or induction at the beginning of their career, and then they practise for the rest of their lives what they have been taught. Once they leave the university, the training college, the apprenticeship, they know all they are ever likely to need to know, or at least what they do not know they can pick up on the job.

In contrast to these well-established views, training (broadly understood) is the single most important factor in the development of excellent companies.

Total Service Quality cannot be achieved without substantial investment in training prolonged over many years.

In *Thriving on Chaos* Tom Peters emphasizes ten points in respect of training.

1. Start as you mean to go on: Immediately an employee starts in the firm, on day one, he or she should be given induction training. The purpose of this is to explain the way the company works, the products the company makes, something of the culture and the operating systems. Sony makes new employees learn the prices and attributes of all its product lines, sales distributions and values, as well as key balance sheet values. The lesson is: start as you mean to go on.

2. All employees should be treated as career employees: Do not treat any group of workers as temporary, and so not worth the trouble. It will hardly appease an irate customer to be told that her problem was caused by a temporary worker whom you did not bother to train fully because he was not going to stay! Everyone who works for a quality company has to have exactly the same standard.

3. Retrain, retrain, retrain: Everyone, from the MD to the cleaning lady, should be regularly brought back to the training room for updates, changes and refreshers. The culture must be that of a sports team: professional players do not stop training when they get onto the team, they train more and harder. They go over the moves and the plays again and again.

4. Spend money and time generously: It is difficult to lay down a rule for calculating the training budget. Most Irish companies, however, could comfortably double what they are spending at present and hardly notice the effect on the P & L account. Training should be perceived as a key strategic area for the company's development. The following steps might be taken:

- set out the objectives of the programme.
- employ a good consultant to initiate the programme.
- train a senior executive to drive the training function.
- establish a detailed training plan based on the operational plan of the company and the skill/knowledge/attribute profile of the employees.
- prepare training programmes and materials to carry out the plan.
- implement and monitor the plan.
- record all training activities undertaken by each employee.
- assess the results.
- review the overall effectiveness of the programme.

5. Organize on-the-job training properly: One of the most valuable forms of training is on-the-job, often derisively called 'sitting with Nelly', *if it is properly organized.* In practice, for specialized functions, this may be the only available way to develop and refine skills. On the other hand, training does not just happen. Supervisors and other specialists have to be allowed the time, and given the incentive and the authority to impart their knowledge and skills. This may mean effectively overstaffing your department, as well as training these people in the best ways of imparting their knowledge.

6. Do not limit what you teach people: If you take a narrow view of what your employees should know, so will they. This is not to suggest that you should establish classes in pottery or calligraphy, but there is an enormous range of business-oriented skills and information that can usefully be imparted. In a simple sandwich bar, for instance, the servers would benefit from a knowledge of hygiene, detailed information about the different kinds of bread and fillings, psychology to help them deal with impatient customers, and statistics to help with problem solving. The more the workforce understands about the business they are in, the nearer you will be to achieving excellence.

7. Training emphasizes management's commitment: Talk is cheap. It is easy for the Group Managing Director to stand on a chair in the canteen and spout about excellence and the quest for quality. Only when specific individual training is applied to each worker does the message sink home.

8. In hard times train more, not less: Training is not a luxury; it is a necessity, like oil in a machine. In hard times, it is doubly important to ensure that the engine runs smoothly. It is during these times, as the PIMS data shows, that quality really provides dividends. So training budgets should be stiffened at these times. It is too easy for management and line supervisors to put off training: 'We're stretched to breaking point, as it is', they say. 'Just wait until we have the current rush over, and then we'll sit down and work out the training programme.' Of course the current rush will never be fully over, partly because the process is continually operated by ill-trained people.

9. Training is a line function, not a staff one: Training must at all times be plugged directly into the needs of the line operators. Ideally it should be initiated and driven by line supervisors and managers. However it is managed, the training programme must be 'owned' by the line. A programme imposed by enthusiastic staffers simply will not stick. The culture change implicit in making line supervisors and managers training oriented is perhaps one of the hardest culture changes of the whole Total Service Quality programme.

10. Use training to change the culture: Training programmes fall into the following categories:

- induction training.
- new skill training.
- skill refreshers/new aspects.
- quality system training.
- problem solving.
- management training.

CASE STUDY: A STAFF VERSATILITY CHART

NAME	YEARS SERVED	QUALITY INSPECTION	STORES MANAGEMENT	COMPUTER KNOWLEDGE	ADMINISTRATION	SALES	PRODUCT KNOWLEDGE	DOCUMENTS
	15	Fully competent	Fully competent	Fully competent	Fully competent	Fully competent	Fully competent	Fully competent
	13	Some experience	Fully competent	Fully competent	Fully competent	Fully competent	Fully competent	Fully competent
	15	Basic knowledge	Basic knowledge	Fully competent	Fully competent	Basic knowledge	Fully competent	Fully competent
	11	Fully competent	Fully competent	Basic knowledge	Fully competent	Fully competent	Fully competent	Fully competent
	1	Basic knowledge	Basic knowledge	Competent under supervision	Fully competent	Basic knowledge	Basic knowledge	Fully competent
	1	Some experience	Fully competent	Basic knowledge	Fully competent	Fully competent	Fully competent	Fully competent
	6 mths	Some experience	Fully competent	Basic knowledge	Fully competent	Competent under supervision	Competent under supervision	Fully competent

Source: A.K. O'Hare

In each of these categories, the company has a chance to inculcate the cultural values and the objectives of the company into the workforce. The training programme is one place where shared values can be created and discussed. It is this direct contact with the employees that makes the training programme such a potentially powerful tool in the campaign to achieve Total Service Quality.

Like other company activities, training should be planned. At the beginning of the year the training needs of each member of staff, department by department, should be identified. Typically the head of training draws up a special form for each departmental head to fill in. This is called the Identification of Training Needs (ITN) form. It lists each member of the department, and the department head, in consultation with the individual, perhaps at the performance appraisal session, identifies training requirements. These might be under various headings, such as 'technical knowledge/skill', 'market/customer knowledge', 'quality systems' etc. The finished forms are then combined by the head of training into a single company-wide training plan for the year.

Performance appraisal

Training should go hand-in-hand with objective performance appraisal. As in other areas of quality management, targets and objectives should be set, and regular progress checks made. Progress reviews provide an opportunity for feedback between the manager and the job holder. This allows confusions and misunderstandings to be cleared up, and opinions to be shared and corrective actions taken where necessary. The appraisal process has three main purposes. These are:

- to measure how the job holder is contributing to the business results (this obviously has implications for rewards).
- to identify how that contribution can be improved.
- to establish a formal channel of communication and understanding between the job holder and the manager.

Performance appraisal should be done for each member of staff at least once a year. The appraisal is a two-way discussion between a manager and a staff member during which they review the staff member's performance in the previous period and make plans to develop and improve performance in the next. A checklist should be given to the staff member at least a week before the appraisal interview. The completed list should be private to the job holder. This will encourage the person being appraised to organize their thoughts honestly, adequately and in detail.

The appraisal interview would normally follow the direction of the checklist. A typical interview might run as follows:

1. Confirm the key work goals. (These should generally be those laid down at the end of the previous appraisal interview.)

2. Discuss how well they were achieved.
3. Explore what helped the achievements and what hindered.
4. Identify strong and weak performance factors.
5. Identify which parts of the work the staff person finds attractive and stimulating and which parts less so.
6. Examine possible changes and improvements in skills, knowledge, systems or operations that might improve the output.
7. Establish the key work goals for the next appraisal period.

At the heart of the appraisal are the performance factors, which have been identified above as a key part of the recruitment stage. During the appraisal interview the manager and the job holder should ensure that their understanding of the significance of these factors is the same.

The results of the appraisal should be recorded by the manager and a copy of the appraisal given to the job holder. This, as we have seen, forms the basis for the next appraisal interview.

ACTION

- **is your staff resource in balance with demand?**
- **review your staff planning and recruitment practices.**
- **how successful were your last few appointments?**
- **how many days per worker do you spend in training?**
- **complete (or review) your Investigation of Training Needs form.**
- **does your performance appraisal system work well? How do you know?**

SUMMARY

People are the key resource in all service operations; yet often less attention is paid to recruiting, training, developing and motivating them than the most basic piece of machinery. This chapter explored ways in which service companies can establish properly based staff planning, recruitment, training and personnel appraisal schemes.

CHAPTER 7

Suppliers

'No man is an island, entyre unto itself . . .'
John Donne

Every business, whether it be mining or agriculture, retailing, heavy goods manufacture, hairdressing or computer programming, depends on customers certainly, but also on suppliers. Even a professional occupation such as the law is dependent on suppliers. Solicitors buy stationery and paper, but also law books, law reports, various research services, legal opinions and representation from barristers, perhaps tax advice from accountants, not to mention equipment. The quality of the final product is dependent on the quality of the purchased supplies. If a law partnership bases its advice on an out-of-date copy of *Every Man's Home Lawyer*, backed up by counsel's opinion from the newest recruit to the bar, however effective they may be in themselves, their procurement policy will let them down. As in other aspects of service quality, customers are not very tolerant of failure. If I buy an answering machine from a supplier and it is faulty, they can replace it; if the second one is also faulty, it is not quite enough for the sales clerk to say 'There's nothing we can do about it . . .' I certainly will not go back there, for anything.

Certain service industry sectors have always been conscious of the need for quality in their purchases. Solicitors take great care to brief top barristers, hospitals tend to buy only the best supplies, advertising agencies look for top quality print. The stories of top chefs insisting on going themselves to the market to buy produce are legendary. Less costly establishments such as hamburger joints do not often take the same trouble. McDonald's, as usual, is one of the exceptions. They recognized from the early days that the quality of the hamburger was almost totally dependent on the quality of the meat, which was extremely variable.

The only regulations relating to hamburger meat at the time controlled fat content at no more than 33 per cent. This did not prevent suppliers adding excess blood, soy protein, water, nitrates (to keep stale meat pink), tripe, and even ground bone to the patties. No wonder hamburgers seethed with water, or hissed with gas when cooked, or bowed on the grill, or squirted like a grapefruit when bitten into, or leaked reddish dye. McDonald's hamburgers were to consist of 20 per cent or less fat, 83 per cent lean shoulder from grass-fed cattle and 17 per cent choice plates (lower rib cage) from grain-fed cattle. The problem was to enforce these standards.

At this time McDonald's used only fresh meat, so they had a chain of 175 suppliers across the States. They drew up a fifty-item checklist for franchisers to use on deliveries, including a hydrochloric acid test for fat content. Suppliers still tried to cheat, so McDonald's insisted on the right to inspect premises at any time, day or night. On one 3.00 a.m. visit they found stale meat being recycled, on another a box of soy powder under a sink. The two suppliers were struck off the list (Love 1988, 129–32).

Few service companies have the weight and persistence to go as far as McDonald's. Yet in any environment the quality of input is critical: as the computer people say, 'garbage in, garbage out'. But how is this to be done?

The first step is to plan the purchasing function, not simply let it happen. Control of purchasing should be documented in a separate procedure, which lays down in some detail the various parameters of the function. In particular, the purchasing procedure must:

- assign responsibility for and within the purchasing function.
- define the manner in which suppliers are selected to ensure that they are continually capable of supplying the requirements.
- indicate the written orders, specifications and certificates of conformity appropriate for each class of purchase (Oakland 1989, 97–100).

The ISO 9004 guidelines state that 'Purchased products and services may be critical to the quality, cost, efficiency and safety of the services supplied by a service organization. Procurement of products and services should be given the same level of planning, control and verification as the other internal activities. The service organization should establish a working relationship with sub-suppliers including feedback. In this way a programme of continuing quality improvements can be supported and quality disputes avoided or settled quickly. Procurement requirements should include as a minimum:

- purchase orders, whether set out as descriptions or specifications.
- selection of qualified subcontractors.
- agreement on quality assurance and verification methods.
- provision for settlement of quality disputes.
- incoming product and service controls (including traceability).
- incoming product and service quality records.

'In considering a subcontractor, the service organization should consider:

- on-site assessment and evaluation of the subcontractor's capability and/or quality system elements needed for quality assurance
- evaluation of subcontractor's samples
- past history with the selected subcontractor and similar subcontractor
- test results of similar subcontractors
- experience of other users.'

1. Specification

In the design of the quality system, the service delivery specification (see Chapter 4) will detail the quality requirements for all purchases. In practice, specifications are written only for goods and services that directly affect the customer's perceived quality. No specifications need to be drawn up for the office pencils or toilet paper, for instance.

Service companies purchase both products and services. Specifications for products are considerably easier to draw up than for services. For products, it is practical to draw up a full description of the type, style or grade required, with detailed technical specifications such as measurements and tolerances. Other items in the specification should include delivery lead times, packing and transport specifications and accompanying documents. Any measurements should as far as possible be traceable back to international standards. Thus products bought by weight should be weighed before delivery on a set of scales calibrated by an approved authority against national standards, and the national standards calibrated against international ones.

Measurements apply to quantity, quality and timeliness. If possible a certificate of conformance, by which the manufacturer declares that the product delivered meets specification, should be presented with the delivery. In many cases this may not be practical. A brand name, backed by a written specification, constitutes a *de facto* declaration to you that the product meets the specification.

Although services are difficult to quantify in this way, they are not impossible. The techniques described in this part of the book are all potentially useful for establishing a proper specification for services. Service specifications should detail the type of service required, the conditions of delivery, the degree of reliability, the service delivery specifications (see Chapter 4), the quality standards as well as requirements such as reliability, flexibility and so on. Of course, the kind of detailed specification possible for a widget may not be practical for services such as those of a barrister, a management consultant or a computer programmer.

2. Selection

Once the organization has decided what it needs, the next step is to identify where this requirement can be met. When designing the service, the service designer and his or her clients must consider which products are to be supplied in-house and which purchased outside the company. This is called the make/buy decision. It crops up in all organizations at several levels. Is our sandwich bar better off buying bread outside, or baking the bread itself? Should an insurance firm employ solicitors or use an independent partnership? Should an accountancy department do the payroll itself, or employ an external specialist company to do it?

Clearly, if all other things were equal, this decision would be a simple one of relative cost. Unfortunately life is unlikely to be so simple. Key non-cost criteria are:

- product quality, including basic specification and consistency.
- reliability and ability to deliver on time every time.
- flexibility and responsiveness to changing demands.
- quality of supplier management.
- good relations with supplier personnel.

Not surprisingly, these are the same five criteria (discussed in previous chapters) by which your customers judge *you*!

Make/buy decisions are complex and crucial. For instance, certain supplies may be construed as critical to the business, while others are less so; there may be only one or at best a few suppliers of critical inputs; in an extreme case the total viability of the business may depend on the goodwill of a particular supplier. The managers of the company may well wish to avoid this degree of vulnerability. The situation should continuously be re-evaluated in the light of changing circumstances. This re-evaluation should not however put excessive weight on short-term financial considerations; in any supplier/customer dealings, a relationship built up over years has immense value.

If the product is to be bought rather than made, a supplier must be chosen. Three considerations arise:

Can the potential supplier make and deliver the product to the specifications?

Can the potential supplier do this reliably?

Can the potential supplier do this at a price we can afford?

The first question is the easiest to answer. Samples of the product or activity should be purchased and tested. In the case of a product, this is a simple purchase and inspection, perhaps to destruction. In the case of a service it is normal to try out the supplier on a small service first (a small print job, or a small law case for instance). This test process should be seen through to the end. Too often in the excitement of establishing a new relationship, repeat orders are piled on a new supplier before the first test has been delivered and assessed. The rule is: do not give the second order until the first has been seen and approved. As we have seen above, the specifications should be documented as far as possible, and copies held by the supplier and the purchaser.

The second and third questions, relating to reliability and price, are somewhat harder to answer. The strategic considerations in choosing a supplier will include an assessment of the criticality of the product purchased, and the proportion of the supplier's business represented by your business. (Too low and you risk being pushed aside for larger customers, too high and the whole business depends on you.) In the past it was considered prudent to have a number of suppliers for each important delivery, thus ensuring continuity of supply and price competitiveness. The modern trend is to reduce suppliers to as few as possible, and bind them closely to the purchasing business. During the 1980s many multinationals reduced their vendor base by as much as 85 per cent.

The amount of time and effort put into assessing suppliers depends on the criticality of the product supplied. An ordinary office is hardly going to spend much effort worrying about suppliers of ball-point pens or jotting pads; furniture and floppy disks merit some more attention, and software and other critical supplies a great deal more again. The following techniques are commonly used to assess suppliers.

Quality questionnaire: A more or less elaborate questionnaire can reveal a great deal about the proposed supplier (see panel for a sample questionnaire). Questions such as:

- do you have a documented quality system, including a quality manual?
- is there a written specification for the finished product?
- what in-process and end-process checks and inspections are in place?

These immediately show the degree of system management the supplier has in place. (As with all the ideas described in this book, this technique is not appropriate in all cases. The eminent barrister Mr Buzfuz SC would not take kindly to such a questionnaire, even from the senior partner of Dodson and Fogg!)

Plant visits: In assessing a supplier a plant visit by one or more of your executives is desirable, particularly for one with whom it is hoped to develop a significant long-term relationship, or one whose input is critical. This should not be a mere stroll, but a serious attempt to explore the capabilities of workers and management. It may not be practical to do this for overseas suppliers.

Auditing: An audit is a formal version of a plant visit. It is a systematic attempt to explore the supplier's operational systems, based on a checklist and a questionnaire. This questionnaire should normally be based on a recognized quality system such as ISO 9000. Of course, fulfilling requirements and completing documentation in a mechanical way is a necessary but not sufficient condition for the production of continuously improving quality. The auditor should also use his or her eyes and ears to evaluate what they see and hear about them during the visit.

CASE STUDY: CALCULATING THE COST OF DISSATISFIED CUSTOMERS

Cost of Poor Service Quality Calculator Worksheet

Affiliate ______________
Department ______________

LOST REVENUE

A. Annual Revenue $ ______________

B. Number of Customers ______________

C. Percentage of Dissatisfied Customers ______________

D. Number of Dissatisfied Customers (C x B) = ______________

E. Percentage of Dissatisfied Customers Likely To Switch (60–90%) × ______________

F. Number of Dissatisfied Customers Who Will Switch (D x E) = = ______________

G. Revenue Per Customer (A ÷ B) = $ ______________

H. Revenue Lost Through Poor Service Quality (F x G) = $ ______________

EXCESS EXPENSE

A. Annual Operating Expense $ ______________

B. Percent of Expense To Do Things Over Again and Placate Dissatisfied Customers (25–40%) × ______________

C. Excess Expense Due To Poor Service Quality (A x B) = ______________

Source: Bank One

During the audit a supplier can be scored on various critical elements of the quality management system. The Malcolm Baldrige auditors for instance divide their scoring into three evaluation dimensions, because a firm's planning may be better than its action, and the action may not alas give quite the desired result. The categories are called Approach, Deployment and Results. They score each dimension out of 100 per cent, with 0 per cent representing a lack of system, and merely anecdotal results, and 100 per cent a fully planned and integrated system, actively supported, giving a proven world-class result.

Based on the results from this scoring system, the supplier can be characterized. A typical characterization would be:

- class A: Excellent systems, ship to stock supplier (no incoming inspection required).
- class B: Adequate vendor, but system not good enough to allow product in without incoming inspection.
- class C: Inadequate controls, extreme care needed with product if it must be used.

Third party registration: A substitute for audits by the company itself is third party registration, whereby the supplier's quality management systems have been registered as meeting some standard such as the ISO 9000 series, or the Quality Mark. At least the supplier's system for producing products has managed to meet a consistent standard. Whether that standard is sufficient for your needs is quite another question.

Other assessment methods: Ideally, historical data of the supplier's performance over a number of years would be included in the assessment of a new supplier; of course this is by far the best way of judging likely future performance. Unfortunately it is unlikely to be made available. Trade reputation is the weakest method of assessment. These reputations are often highly subjective, based on little evidence, and subject to rapid revision. Only if the supplier also supplies a purchaser of objectively known high standards can this evidence be at all acceptable.

3. The quality of incoming products

The purchasing process has failed if the product does not meet the specification. The process of ensuring conformance starts with the purchasing company defining its specification in detail. An evaluation of the capability of the supplier's processes forms the second level of assurance. A good system will not, however, rely solely on these initial checks; maintenance of the standard of conformance must be checked regularly. Since the quality of the outgoing product depends on the quality of the supplies, a continuous *formal* monitoring procedure must be established. It is tempting to omit the formal documentation of quality results, especially in a service environment. This is a mistake.

Good management and control depends on good information; it is too easy for the immediate circumstances of the case to bias opinion in one way or another. Thus, for instance, in August and September the managers of a company might be particularly conscious of holidays as a cause of excess overtime: they might decide that if they control the simultaneous taking of holidays more tightly, overtime would reduce drastically. But if the data is properly analyzed over several months, it may perhaps be seen that although holidays are an important cause of overtime in August, they hardly impact at all on other months. The holiday issue was dramatically evident in August, and vividly remembered in September, but it was not the root cause. Action taken on holidays would simply arouse staff resentment without addressing the problem.

Similarly with suppliers: every company goes through a bad patch from time to time, where for instance incident after incident of bad product or bad documentation occurs in a short period. This naturally generates a good deal of exasperation, and an otherwise good supplier is quickly condemned by anecdotal information. A record system should cool tempers (or conversely justify them) by putting a perspective on the problem. Thus although there may have been outbreaks of sloppy work on the supplier's part, on average, over several months, their product is much better than the rival's who delivers a continuous drip of errors. The records enable the purchaser to approach the supplier with information to help him to solve the problem. Perhaps it is as simple as identifying when, for holiday or sickness reasons, the warehouseman *and* his assistant are away.

Records of performance can be very simple, a sort of 'Customer Comment Card' detailing whether a particular service was on time, efficient, met flexibility needs etc., whether the product was of good quality, the right quantity and so on, whether the documentation was accurate and easily understood. These records must be analyzed subsequently and regularly. Price's three rules of quality control should be remembered in this context:

- no inspection or measurement without proper recording.
- no recording without analysis.
- no analysis without action.

(Price 1984, 29)

Inspection: To maintain the quality of incoming products, some kind of inspection of the product is likely to be necessary. This may be on a 100 per cent basis, or more usually on a sample basis. Normally, the first lots delivered by a new supplier receive special attention, and this attention is usually repeated when an established supplier delivers a new product.

Typically in these cases a 'first item' sample is delivered before the main consignment, and this first item is given detailed and scrupulous attention. Assuming it passes this test, a signed or authorised 'Approved Item' is returned to the supplier to act hereafter as the touchstone for future deliveries. A counterpart of this is kept by the purchaser.

When the bulk delivery takes place, the incoming product is tagged or otherwise identified. This will prevent it being used until it has been approved. Usually a random sample of the batch is taken and evaluated. Assuming the product passes the prescribed tests and inspections, the batch tags are changed, and the product can be used. If the product fails to pass all the tests, then it should be re-tagged and removed from possible use, ideally into some kind of quarantine area. The effectiveness of the incoming inspection routine is dependent on the accuracy and discipline of the tagging/identifying process. There is no point in an inspector elaborately examining an incoming product already being processed; if this is allowed to happen, not only is the incoming inspection routine rendered a farce, but the whole quality of the product you are making is put at risk.

This of course applies throughout the process: non-conforming products must be identified and removed from the process in such a way as to prevent their slipping back into use. This principle has wide application. For instance, to ensure its quality of service, a travel agency must constantly update its time-tables, *and* discard the old ones, lest they be used accidentally. Accountants must have a system in place to ensure that all its audit clerks are conforming to the latest accounting standards. Yesterday's stale buns must be effectively identified and removed from the tea-shop counters; stockbrokers must quote and use the very latest prices; lawyers the latest judgments, even if the Supreme Court judgment has not yet been published and the one from the High Court has.

The statistical regulation of inspection is one of the most elaborate mechanisms of quality control. In this country it is usual to use the protocol devised for the US Department of Defense called MIL-STD-105E, which was published as a revision of MIL-STD-105D in May 1989. The purpose of this protocol is to provide a decision table by which the results from a sample can be extrapolated to the batch as a whole. Thus, assuming I am looking for a certain quality level, how can I know from my relatively small sample whether the supplier has achieved this?

The system starts with two assumptions:

- The sample is drawn at random from the batch (random in this context means that every item in the batch has an equal chance of being selected).
- The batch is made of articles made under the same conditions.

Once these conditions are met, the protocol is based on various tables. These tell the inspector what sample size to take for various batch sizes. Thus in the normal inspection of a batch of 1,000 items, the specified sample may be 80 items (based on the usual starting point of General Inspection Level II). These are taken at random from the batch and examined. If the specified Acceptable Quality Level is set at 1 per cent, then if more than 2 defective items are found in the sample of 80, the whole batch will be rejected as being below quality.

TABLE I – SAMPLE SIZE CODE LETTERS

Lot or batch size			Special inspection levels				General inspection levels		
			S-1	S-2	S-3	S-4	I	II	III
2	to	8	A	A	A	A	A	A	B
9	to	15	A	A	A	A	A	B	C
16	to	25	A	A	B	B	B	C	D
26	to	50	A	B	B	C	C	D	E
51	to	90	B	B	C	C	C	E	F
91	to	150	B	B	C	D	D	F	G
151	to	280	B	C	D	E	E	G	H
281	to	500	B	C	D	E	E	G	J
501	to	1200	C	C	E	F	G	J	K
1201	to	3200	C	D	E	G	H	K	L
3201	to	10000	C	D	F	G	J	L	M
10001	to	35000	C	D	F	H	K	M	N
35001	to	150000	D	E	G	J	L	N	P
150001	to	500000	D	E	G	J	M	P	Q
500001	and	over	D	E	H	K	N	Q	R

TABLE II – Single sampling plans for normal inspection (Master table)

Acceptable Quality Levels (normal inspection)

Sample size code letter	Sample size	0.010	0.015	0.025	0.040	0.065	0.10	0.15	0.25	0.40	0.65	1.0	1.5	2.5	4.0	6.5	10	15	25	40	65	100	150	250	400	650	1000
		Ac Re	Ac Re	Ac Re	Ac Re	Ac Re	Ac Re	Ac Re	Ac Re	Ac Re	Ac Re	Ac Re	Ac Re	Ac Re	Ac Re	Ac Re	Ac Re	Ac Re	Ac Re	Ac Re	Ac Re	Ac Re	Ac Re	Ac Re	Ac Re	Ac Re	Ac Re
A	2	↓	↓	↓	↓	↓	↓	↓	↓	↓	↓	↓	↓	↓	↓	0 1	↓	↓	1 2	2 3	3 4	5 6	7 8	10 11	14 15	21 22	30 31
B	3	↓	↓	↓	↓	↓	↓	↓	↓	↓	↓	↓	↓	↓	0 1	↑	↓	1 2	2 3	3 4	5 6	7 8	10 11	14 15	21 22	30 31	44 45
C	5	↓	↓	↓	↓	↓	↓	↓	↓	↓	↓	↓	↓	0 1	↑	↓	1 2	2 3	3 4	5 6	7 8	10 11	14 15	21 22	30 31	44 45	↑
D	8	↓	↓	↓	↓	↓	↓	↓	↓	↓	↓	↓	0 1	↑	↓	1 2	2 3	3 4	5 6	7 8	10 11	14 15	21 22	30 31	44 45	↑	↑
E	13	↓	↓	↓	↓	↓	↓	↓	↓	↓	↓	0 1	↑	↓	1 2	2 3	3 4	5 6	7 8	10 11	14 15	21 22	30 31	44 45	↑	↑	↑
F	20	↓	↓	↓	↓	↓	↓	↓	↓	↓	0 1	↑	↓	1 2	2 3	3 4	5 6	7 8	10 11	14 15	21 22	↑	↑	↑	↑	↑	↑
G	32	↓	↓	↓	↓	↓	↓	↓	↓	0 1	↑	↓	1 2	2 3	3 4	5 6	7 8	10 11	14 15	21 22	↑	↑	↑	↑	↑	↑	↑
H	50	↓	↓	↓	↓	↓	↓	↓	0 1	↑	↓	1 2	2 3	3 4	5 6	7 8	10 11	14 15	21 22	↑	↑	↑	↑	↑	↑	↑	↑
J	80	↓	↓	↓	↓	↓	↓	0 1	↑	↓	1 2	2 3	3 4	5 6	7 8	10 11	14 15	21 22	↑	↑	↑	↑	↑	↑	↑	↑	↑
K	125	↓	↓	↓	↓	↓	0 1	↑	↓	1 2	2 3	3 4	5 6	7 8	10 11	14 15	21 22	↑	↑	↑	↑	↑	↑	↑	↑	↑	↑
L	200	↓	↓	↓	↓	0 1	↑	↓	1 2	2 3	3 4	5 6	7 8	10 11	14 15	21 22	↑	↑	↑	↑	↑	↑	↑	↑	↑	↑	↑
M	315	↓	↓	↓	0 1	↑	↓	1 2	2 3	3 4	5 6	7 8	10 11	14 15	21 22	↑	↑	↑	↑	↑	↑	↑	↑	↑	↑	↑	↑
N	500	↓	↓	0 1	↑	↓	1 2	2 3	3 4	5 6	7 8	10 11	14 15	21 22	↑	↑	↑	↑	↑	↑	↑	↑	↑	↑	↑	↑	↑
P	800	↓	0 1	↑	↓	1 2	2 3	3 4	5 6	7 8	10 11	14 15	21 22	↑	↑	↑	↑	↑	↑	↑	↑	↑	↑	↑	↑	↑	↑
Q	1250	0 1	↑	↓	1 2	2 3	3 4	5 6	7 8	10 11	14 15	21 22	↑	↑	↑	↑	↑	↑	↑	↑	↑	↑	↑	↑	↑	↑	↑
R	2000	↑	↑	1 2	2 3	3 4	5 6	7 8	10 11	14 15	21 22	↑	↑	↑	↑	↑	↑	↑	↑	↑	↑	↑	↑	↑	↑	↑	↑

↓ = Use first sampling plan below arrow. If sample size equals, or exceeds, lot or batch size, do 100 per cent inspection.
↑ = Use first sampling plan above arrow.
Ac = Acceptance number.
Re = Rejection number.

A sampling plan for attributes: products classed as defective or non-defective *US MIL-STD 105E.*
Shading indicates rows/columns for a 1,000 item batch at 1% AQL.

There are a number of sophistications to this simple regime which make the sampling plan extremely flexible. For instance, it is possible to establish inspection at three possible levels: Tightened (for problem suppliers), Normal, and Reduced; a path is laid down for the movement from normal to reduced and back again. Various inspection levels are also possible, again depending on the degree of rigour required. These are broken into General Levels (I, II and III) and Special Levels (S-1 to S-4). General Level III is the most demanding and calls for the largest samples, Special Level (S-1) the least. Unless otherwise required, General Level II is typically used. Double and multiple sampling plans are also included in the standard.

The first and fundamental requirement for the use of these plans is to decide just what proportion of non-conforming items will you accept as reasonable? In theory of course one would like to be able to say simply Zero Defects; in practice this standard, or some such expressed as a small number of parts per million, may be necessary in life-threatening circumstances. In many environments, the extra cost of controls and inspections necessary to ensure this level are simply too costly for the market. The question then is: how can we achieve the highest degree of error-freeness appropriate to the market?

This level is normally expressed as the Acceptable Quality Level, which is the average proportion of defectives acceptable. Since this figure is an average, not all of the results will take the AQL value. Perhaps one in ten deliveries will have a defective proportion well over the average. This is clearly not suitable. In practice therefore the AQL level becomes a limiting level (European Organization for Quality, 1988). Thus if you wish to achieve an *average* incoming quality level of 1 per cent, setting the AQL level at 2.5 per cent would guarantee a 1 per cent Average Outgoing Quality Level.

The sampling tables are designed to establish a reasonable security that certain levels of quality are being maintained. As in any sampling process, the key to the process is the size of the sample, not as a fraction of the batch size, but in relation to the *variation* in the batch. If (as a visiting Martian) we wish to discover how many heads the average human has, we can make a good estimate from a very small fraction of the 5 billion people on earth; the number of heads per human does not vary. If, on the other hand, we wished to make an estimate of religious or political opinions, a much larger sample of the same population would be necessary. Obviously as the sample grows larger, one is entitled to feel more secure in one's estimate; our Martian researcher is likely to have made an extremely accurate guess by counting the heads of the first human it met; on the other hand, the more human heads counted, the more secure becomes the result.

The table below shows the effects of this on the sampling of incoming goods. Suppose you have fixed a 1 per cent AQL level with your supplier, and he actually delivers a batch with 5 per cent defective. The question is what is the likelihood that your sample will reject this batch?

Sampling Results from MIL-STD 105E

Lot size	Sample	Sample %	Probability of acceptance	Defective level for 10% acceptance
500	50	10.0	27	7.56
1,000	80	8.0	23	6.52
2,000	125	6.25	17	5.35
5,000	200	4.0	7	4.64
12,000	315	2.63	2	3.74

Thus, if the batch size is 1,000 items, the tables (Normal Level II) suggest a sample of 80; this is an 8 per cent fraction of the batch. Although the AQL is 1 per cent and the batch is 5 per cent defective there is just under a one in four chance that the batch will, after all, be accepted. The next column of the table shows that one in ten batches with defective levels as high as 6.5 per cent will be accepted. As the sample size grows larger, the discrimination of the sampling plan becomes greater. At the 12,000 batch size, the sample is 315, and there is only a one in fifty chance of accepting the 5 per cent defective batches. Counter-intuitively, the sample fraction of only 2.6 per cent gives a much better result than the sample fraction of 10 per cent.

In deciding with the supplier what AQL level is appropriate, the purchaser needs to examine the effects of various possible things that might be wrong with the supplied goods. Clearly pencils and paper clips will receive less attention than software or food products. For each possible defect, or at least the so-called critical defects, the following questions should be asked:

- how easily is the defect recognized? The more expensive the testing process, the fewer you wish to test and therefore the higher the AQL level.
- where is the defect recognized? The later in the process that the defective product is recognized the more damage it may have done.
- what is the seriousness of the defect to the customers? Minor cosmetic defects are less serious than structural ones.
- what is the cost of rectifying the defect? Replacement cost, delays in establishing new deliveries, other knock-on problems internally, failure to deliver to the customer on time and resulting loss of goodwill, etc.

The answers to these questions, perhaps graded in some form of rating system, will give the purchaser a feel for the level required, as a trade-off against the cost of achieving that level.

ACTION

- **list all purchases: isolate critical and non-critical.**
- **identify purchase criteria/characteristic failure modes for these critical purchases.**
- **establish (or review) the existing specifications for purchases.**
- **establish (or review) performance of existing suppliers.**

SUMMARY

Every business depends on the quality of its suppliers, whether of products or services. The purchasing function should therefore be carefully planned. Specifications for purchases (products and *services) should be drawn up, suppliers carefully selected and their performance monitored, and the quality of incoming products must also be carefully assessed.*

CHAPTER 8

Queues and Schedules

'If all you've got is a hammer, every problem has to be a nail.'
A. Maslow

Every definition of service industry includes timeliness. A service is normally created in 'real time', here and now. You cannot keep a stock of phone-calls, medical operations, banking services, or on-time air flights for dispensing when required. What is more, for virtually every service, customers want timely (if not immediate) service, regardless of the simultaneous arrival of other claimants to the same service.

There are exceptions even to this rule. In the early 1980s a US bank decided to give its customers shorter queues. After some work, they began to measure queuing time, and eventually managed to reduce the average queue time from (say) 2 minutes to 30 seconds. Delighted with themselves, they commissioned a market research company to check the customers' reactions. To their dismay the response was not favourable, particularly for two types of customer. For one group, the advertisements announcing the new departure had raised expectations so high that even a 30 second average waiting was regarded as a breach of promise. (A gap had been opened up between these customers' expectations, created by the company itself, and the actual performance.) Another group, senior citizens, regarded dropping in to the bank and having a bit of a chat in the queue as a pleasant part of the day, and they were not happy in having this curtailed. (In this case the gap was between customers' actual requirements and management's perception of those needs.)

Whatever the objectives of the people in the queue, the question of how to design and manage queuing systems and schedules is of prime importance to all service operations. The mathematics of queues and schedules has in fact formed a staple part of the armoury of operations research practitioners since the early part of the century, when the first models of queues were derived. Yet despite this large body of work (most of it admittedly related to a manufacturing context), these concepts have received little attention in the many books published on service quality.

In a sample of the sources used for this book, the words 'queue' and 'schedule' were discussed in 10 pages out of 3,400. The only authors who referred to scheduling at all, Gillian and Bill Hollins's *Total Design* set a strong note: their chapter on the subject is called 'Queuing — the sign of a bad service'. The reluctance to address this key issue is perhaps because the concepts used in queuing and scheduling theory are relatively technical even though it is a well-developed body of work. This chapter will undoubtedly use more mathematical concepts than the whole of the rest of the book put together.

The first step for the designer of any service is to realize that:

1. Queues are *not* inevitable, and
2. 'First-come-first-served' is not the only or even the best method of arranging a queue.

The length of any queue depends on the number of service points and the speed of service. The more service points, the less the queue. The very existence of a queue means that the service designer has weighed the cost of an extra service point against the inconvenience to the customers in having to wait, and the customers lost. Put like that, you wonder that people tolerate queues at all!

However, since infinite service capacity is not a practical proposition, there will be queues. If there are queues of people or tasks waiting to be serviced, a question arises as to which should be tackled first. Imagine a case where you have five tasks for clients waiting in the 'in' tray. You can take these in any order you like, as long as they are all completed by the end of the day. There are in fact as many as 120 possible orders to choose from. (There are five to choose from for the first task, which leaves four to choose from for the second, three for the third, and so on.)

If you had the relatively modest number of ten jobs to complete in a week, the same arithmetic says there are more than 3.6 *million* different orders in which these tasks might be taken! It is clearly impractical to examine each one of the possible orders of tasks to see how well they meet operational objectives before starting work. Does it matter? Is there likely to be any benefit in choosing one of these orders rather than another? If there is no benefit, clearly first-come-first-served is as good as any. If however benefit to the customers or the firm (or both) can be derived from adopting some non-random order, then it behoves a service designer to examine how this might be done.

Scheduling and queuing are two sides of the same problem: how to deliver a service as efficiently as possible. There are two key problems:

- in what order should the various jobs be taken?
- how many service stations should be available?

The first is the problem of sequencing or scheduling, the second a question of resource allocation.

CASE STUDY: TEXACO CUSTOMER CARE PROGRAMME

In early 1989 the Company recognised a need in the market place for a Quality Customer Care Programme.

A test of six pilot sites was created in the greater Dublin area, where the concept of 'Experience the Best Forecourt Service there is' was put to the test. The customer feedback indicated clearly that we were on the right track and that service was here to stay.

Under the guidance of Managing Director, Vincent O'Brien, the people concerned were brought together and he formed a team to spearhead the programme. This team consisted mainly of Field people who researched and developed the initial stages of the Customer Care Programme.

Stage one was to create an awareness with our customer that doing business with Texaco would be the nicest experience a customer could imagine. This was done by means of advertising, using TV, Radio, outdoor posters, and weekend events. We also reinforced the concept of Customer Care with our top retailers nationwide, and their feedback was documented and analysed by our Field people. This helped the team to fine-tune the direction of Customer Care and allow them to progress the programme to a higher plateau. This stage saw the introduction of our 'Say Hello to Texaco' Campaign.

In early 1990, the team and our Advertising Agency developed the second generation of our Customer Care Programme when they produced the now famous 'we're putting the service back into service station', and this became our mission for 1990. The outdoor posters mirror imaged the TV/Radio advertisements and brought home the message that we were serious about this service business. But what did service mean — different things to different people. Most of all it meant friendly people, a nice place to do business and equipment and facilities that worked first time and every time.

We again re-trained our retailers and front line troops in the concept of Quality and Customer Care. Some of the areas we focused on were the need for uniforms, complaint handling, cleanliness, and equipment maintenance. Our message to them was that Customer Care Management gives added value in today's competitive environment, with tight retail margins. It was asked how can you provide added value without incurring additional cost? Part of the solution is back to basics — 'service with a smile and it pays'. In Texaco there is continuous monitoring of the Programme to ensure that efforts are producing happy satisfied customers. It was rewarding to see that during the National Marketing Conference of October 1990, Texaco Ireland's Customer Care Programme featured as a major focus point on the agenda, when the Managing Director of Lansdowne Market Research, Robin Addis, cited Texaco's commitment to their current Customer Care Programme. Results of his independent survey were that 'encouragingly, Texaco's Customer Care strategy is showing dividends'.

Source: Texaco

Queue discipline

In environments where the public is being served directly, there is little point in discussing the question of who should be served first. Hardly any system is generally seen to be fair apart from first-come-first-served (otherwise known as First In, First Out — FIFO). Indeed few actions evoke such instant aggression as queue jumping, particularly on the road.

In environments where the work is done in offices, out of the general sight, a more sophisticated approach can be taken. In a non-public situation, the management of the queue can produce a more rational approach, to maximize general benefits. You and the customers have certain requirements for economy and efficiency. In particular the job scheduling system may bring three decision-goals into the assessment. The company may wish, for instance, to maximize:

- resource usage.
- rapid response to demand.
- close conformance to deadlines.

One or all of these objectives may be maximized, within the constraint of a certain cost or resource allocation. This is the first limitation on sequencing; there are also likely to be practical limitations on the order in which tasks are performed. Letters must be written and envelopes prepared before they can be sorted; anaesthetic given before operations can be performed, sandwiches made before cash is received, and so on, even though this may lead to less than optimal resource allocation.

To meet this need, certain rules have been evolved by writers in sequencing and scheduling theory (see for instance Baker 1974). The basic rules have been evolved for the relatively simple single process schedule, where many jobs are queuing to be processed by this operation. In these models a process may be something as simple as buying a ticket from a railway clerk or as complex as insuring an oil tanker. A 'process' may be defined at large as an entire office, or in small as the activities of a single clerk. Someone has to decide in which order the jobs should be taken. The rules by which this decision is made are called the dispatching rules. (A key advantage of the dispatching rule called first-come-first-served is that it administers itself.)

The most important of these rules takes the jobs in order of expected processing time. This is called the Shortest Processing Time system, or SPT. Although this seems 'unfair', it is quite logical, and leads to the counter-intuitive result that first-come-first-served systems are not the most efficient, or even the fairest that might be devised. In fact,

Queuing systems which order jobs by length of processing time will have significantly lower average throughput times, fewer people in the queue and less 'lateness' than any other possible sequencing system.

[Average 'lateness' in queuing jargon means the combination of early and late deliveries. If no credit is to be taken for early deliveries, the term used is 'tardiness'.]

This is the fundamental theorem of sequencing theory. Every other method of sequencing jobs is normally measured against the result obtainable by this method. Only in a few special circumstances does any other rule produce a better result.

The calculation works as follows: time spent in any system depends on a combination of time spent *waiting* to be processed, and time *being* processed. From the customers' point of view the whole time in the system is important, not only when his or her particular problem is receiving attention. The analogy with air flight is good. To fly from Dublin to London takes just under an hour; airlines tend to boast of their achievement of high rates of punctuality. Yet to travel to central London from, say, Sandymount takes much longer. You must allow about three-quarters of an hour by car to the airport, three-quarters of an hour to check in, an hour's flight, at least an hour getting out of Heathrow to a destination in London. The customer's total journey time amounts to three and a half hours, of which two and a half are spent getting into and out of the system. In manufacturing terms, there are two and a half hours make-ready for an hour's production.

Nothing in scheduling theory will affect processing time; if you are your own customer, there is no benefit in ordering a set of personal jobs in this way, since there is no benefit to be derived from the shortening of the queue. However if there is a queue of customers, they can benefit from a reduction. Time spent waiting to start processing (queuing time) is directly dependent on the processing time of those ahead in the queue. If the time they spend being processed is short, so is time in the queue. Consequently the total time in the system will be shorter than usual, although the processing time itself has not changed. If the objective of the company is to reduce the average queue length, and the average queue size or the average lateness, the way to do this is to take all jobs in the order of their processing time, shortest first.

The arithmetic of SPT works is demonstrated in the table, where the results for the SPT scheduling system are compared with the 'fair' first-come-first-served method. Five jobs are in the queue, labelled A to E. Each has a different processing time.

The total processing time is the same for FIFO as SPT, but the total queuing time, and therefore the total time spent in the system (called flowtime), is quite different. Under FIFO the five clients spend an average of 70 minutes in the system, and under SPT an average of 51 minutes, or over 25 per cent less. This gain in efficiency is quite typical.

Comparison of FIFO and SPT scheduling
(Time units may be minutes, hours, days or weeks)

	FIFO		SPT			Total Time in system		
Job Order	Process Time	Queue Time	Job Order	Process Time	Queue Time	Job	FIFO	SPT
A	20	0	D	5	0	A	20	35
B	40	20	C	10	5	B	60	75
C	10	60	A	20	15	C	70	15
D	5	70	B	40	35	D	75	5
E	50	75	E	50	75	E	125	125
Totals	125	225		125	130		350	255
Averages	25	45		25	26		70	51

Not only is the total flowtime much less under SPT, but the queue length is almost always shorter. After sixty units for instance in the FIFO system, only two of the five clients have been processed; under SPT we are more than half-way through the fourth client. Not only is this efficient in terms of space usage, but the customers themselves prefer to see a queue moving, especially if they are at the end of it.

Under SPT the clients as a group and the system overall benefit: who loses? In our example the losers are clients A and B, the first two in the queue. In A's case his time in the system goes from 20 units to 35, and B's goes from 60 units to 75. Naturally clients A and B would not much like this if they knew. On the other hand, given the concept of a client's business with the organisation as a stream of transactions over time, it would be reasonable to point out that another day they might not be first in the queue. Over time SPT will even out the random effect of the queue order. Although the FIFO system is perceived to be the fairest possible, paradoxically it tends to penalize the average customer to the benefit of the untypical (Starr 1989, 574–88).

If one of the clients is regarded as more important than another, it is possible to weight the average process time. For instance, it may be necessary to prevent certain particularly long jobs always being put to the back of the queue. In this event jobs that had been in the queue one period might receive a weighting of 1, those in the queue two periods 2, and so on. The process times are divided by the weights, and the processing order based on the process time divided by the weights. Given the weighting, the SPT sequence of the weighted data is still the most efficient possible.

(A personal time management point occurs here. Although you will not reduce the total amount of work to be done, there may be an advantage in doing the shortest jobs in your in-tray first. By using some derivative of SPT, the jobs can be processed so as to maximize the satisfaction of the various people who are waiting for decisions or results. One's own workload seems more manageable with fewer items in the queue, and at least quick and easy jobs are polished off. The fewer people whose own work is held up looking for inputs the better.)

The schedules discussed so far were not dominated by time considerations. Sequencing theory distinguishes lateness from tardiness. Lateness is defined as completion date minus due date; it may be positive (when a job is late) or negative (when a job is early). Tardiness on the other hand only measures missed deadlines; if a job is on time or early, tardiness equals 0. This is to account for a system where penalties are imposed for missed deadlines, but the system gains no benefits from early delivery.

Surprisingly perhaps, average job lateness is also at a minimum under SPT rules, even though no account is made in SPT of due dates. However, 'average lateness', as technically defined, in effect gives the operation 'credit' for early completion. This may not be sensible. Since SPT systems tend to penalize long jobs, even in the weighted mode, the resulting *tardiness* may be unacceptable under pure SPT.

The normal strategy in this circumstance is to adopt a dispatching rule based on the Earliest Due Date (EDD). Use of this rule in a single process operation will also ensure that the worst case of late completion will be reduced to a minimum. However if you are trying to minimize tardiness in a multiple operation process, it may be desirable to use a rule which takes into account the ratio between the overall amount of work remaining and the amount of time left. This rule looks for the job with the least amount of slack time available between now and the due date.

In a situation where there is a static queue at the beginning of the work session, as for instance where dispatching decisions are made at the beginning of every day or week, and the problem is to complete the work in as short a time as possible, SPT or WSPT (weighted shortest processing time) sequencing is unbeatable, except in the case where the critical objective is to reduce the amount of tardiness. However in real-life jobs often arrive intermittently during the day, and urgent jobs may be allowed to break into the queue. This is called pre-emption.

In service industry environments it is normally possible to break into a job and then resume activity more or less where one stopped. In this case a development of SPT called Shortest Remaining Processing Time (SRPT) should be used. At each decision stage in the day, as new jobs come into the system, the queue should be reformed, giving priority to those jobs with the least amount of work remaining to do. No attempt is made to look forward into the system. Once again, as with SPT, this strategy will reduce to a minimum the average time jobs spend in the system.

These results have been proved by operations researchers working largely in manufacturing environments. Much more complex problems, including those dealing with multiple processes, parallel identical processors, dependent jobs (job A must come before job B, and so on), have been dealt with by extensive simulations to test various combinations of rules in various circumstances. Study after study has proved that only in special circumstances can the ordinary or weighted SPT rule be improved on.

How many service stations?

So far we have discussed ways in which the 'fair' first-come-first-served dispatching rule might be improved upon. Evidently, this can easily be done. However it is difficult to see how any company is going to persuade the ordinary public of the superior rationality of SPT systems. As one author put it, first-come-first-served has an almost sacred status with which companies tamper at their peril. This is best done by those systems where one single queue is served by several service points (check-outs, banks, etc.). In these circumstances no one will get stuck behind particularly slow customers while watching the queues at other tills or counters rapidly disappear. In this environment the questions for management are largely to do with resource allocation: i.e. how many service points should be open at any one time in order to keep queues to an acceptable level? Is the pressure of demand predictable, or do we need a stand-by team?

In special circumstances, a limited form of pre-emption may be acceptable. For instance in a hospital out-patient department, heart failures will be allowed priority over less urgent requirements. In a friendly office environment, shortest job first (up to a point) prevails at the photocopier. Given this constraint, the service industry manager must examine the allocation of resources to the FIFO system to ensure customer satisfaction. In fact considerable operations research work, going back to a paper originally published by a Danish telephone engineer in 1909, has also been done on queuing theory.

Queuing theory distinguishes four basic queue models:

- single server single stage: the basic queue situation.
- multiple server, single stage: one queue served by a number of service points, as in many banks.
- single server, multiple stage: in buying a carpet at a department store, the customer queues for service, then waits (in a theoretical queue) to see the sample, queues again to deliver the buying decision, queues again to have credit checked, queues again for the delivery van.
- multiple stage, multiple server: as above, but with multiple servers at each stage.

Since each of the basic queue situations can be mathematically modelled, managers can examine the effects of adding or reducing the number of stations, changing the number of servers from single to multiple on the queues, or even of redesigning the service to multi-stage rather than single stage. Is it better for a department store to have one queue for service and another for payment, thus reducing the service time at the expense of the total time in the system, or should the server also handle payment? Queuing theory enables the service designer to say what is meant by 'better'.

The theory starts with the idea that both arrivals and service times vary, but both are distributed according to known probability distributions. Typically a model might assume that over a certain period the number of customers arriving will conform to an average. What is not known is the parameter of the distribution. Over a five minute period perhaps 100 customers pass a newsagent's shop; perhaps one in ten turn into the shop and so on average the shopkeeper will receive ten customers per five minutes, or two a minute. This is just comfortable, because each customer takes about half a minute to serve.

However, averages being what they are, people will turn into the shop at random: sometimes several together, sometimes none. The third customer in the queue may therefore have to wait some minutes before being served. Perhaps he or she will leave the shop in disgust to get the newspaper elsewhere (this is called reneging). So as well as the average number of customers per period, the shopkeeper has to attend to the distribution over time. Using the Poisson distribution, we can estimate that if the customers arrive on average two per minute, the chances of three or more customers arriving at once is just over 32 per cent.

Using the Poisson model to estimate business flows*

Average Arrivals	1 per minute	2 per minute	3 per minute
Actual Arrivals (per minute)	Chances of that number (%)	Chances of that number (%)	Chance of that number (%)
0	36.8	13.5	5.0
1	36.8	27.1	14.9
2	18.4	27.1	22.4
3	6.1	18.0 } = 32	22.4
4	1.5	9.0 } = 32	16.8
5	0.3	3.6 } = 32	10.1
6	-	1.2 } = 32	5.2
7	-	0.3 } = 32	2.1
8	-	-	0.8

*Note: The Poisson formula by which these figures are derived is explored in detail in operations research textbooks such as Lapin 1988.

We can see from the table how sensitive the model is to an accurate estimate of the average number of arrivals. If the average arrivals is correctly estimated at 1 per minute (during a lull period), since there is a 37 per cent chance of no customers in that period, the operation will be quiet for considerable periods. In busier times, there is a much smaller chance of no customers, and the numbers of customers arriving together in the shop becomes uncomfortable. This sensitivity to the average means that if the flow of business varies widely, any one Poisson model may only be suitable for limited periods. The resources needed to handle arrivals at one per minute, where there is a less than 10 per cent chance of more than two at once, are much less than those needed to cope with the three a minute average, where there is an 80 per cent chance of more than two, and nearly a 20 per cent chance of five or more. In most queuing situations, arrivals are assumed to follow a Poisson process, with the number of arrivals in any one period having a Poisson distribution. This distribution is described in detail in statistics and operations research texts.

Given these assumptions, the model allows one to analyse various combinations of queues and service points:

- the probability distribution for the number of customers in the system.
- the average number of customers in the system (i.e. queuing and being served).
- the average time customers spend in the queue.
- the average number of customers in the queue.
- the proportion of time a server spends with the customers.

Consider the case of a queue involved in the operation of a central supply room for a large office (taken from Lapin 1988, 474–5). Employees pick up various stationery supplies, served by a full-time clerk. Each request takes an average of 2 minutes to complete, so the clerk can handle 30 internal customers in an hour. About 25 employees call every hour. Following the distribution assumptions noted above, and the convention that average arrival rate is L, and average service rate is M, the following formulae apply.

- Average number of customers in the system (being served or waiting)

$$\frac{L}{M - L} = \frac{25}{(30 - 25)} = 5 \text{ customers.}$$

- Average customer-time spent in the system

$$\frac{1}{M - L} = \frac{1}{(30\text{–}25)} = 12 \text{ minutes.}$$

- average queue length

$$\frac{(L \star L)}{M(M - L)} = 25 \star \frac{25}{(30(30\text{–}25)} = \frac{25}{6} = 4.17 \text{ people.}$$

- server utilization time

$$\frac{L}{M} = \frac{25}{30} = 0.83 \text{ or } 83\%.$$

Using these formulae management can discover that in this system there is little chance of the supply room becoming a social centre: there is only a 10 per cent chance of two employees leaving the centre at the same time. Other interesting results can be obtained from queuing models. One in particular contradicts the common-sense view that two servers will produce identical results to one who is twice as fast. This is clearly not true. For two machines the queue time would be markedly shorter; on the other hand because the process time is much faster, the overall time in the system, the flowtime, is considerably less. The ratio between the queue time and the process time controls which of the two is better.

There are very many subtle variations on queuing models which it would be inappropriate to discuss here. For instance, the model described here is appropriate for situations where arrivals occur singly over time; this is inappropriate for situations where customers arrive all together, such as in rushes for taxis at a railway station. The models will work best over quite short periods of time, because the average arrivals change. The queues at a supermarket check-out will be very different on Friday night than on Tuesday morning.

Rather than explore the details of queuing theory, the service designer may prefer to use a simulation to explore the likely patterns of queues. This technique uses a model of the service environment, based as closely as possible on actual observed figures (average service and waiting times, average and distribution of arrivals, etc.). Using these distributions, and a random number generator (from tables or built into a computer program), it is possible to simulate many months of activity in a single morning. The model is sensitive to its initial assumptions, so it is important to re-run the simulation several times using different assumptions to test the strength of any conclusions you wish to draw.

Various specialist software have been written to handle simulations, the best known being GPSS (General Purpose Simulation System). A micro-version, to run on IBM PCs and Macintoshs is available (Stahl 1990). This simulation program will enable users to run simulations ranging from the flow of customers in a department store, telephone calls to a switchboard, products through a factory, patients through a clinic, documentation through an insurance company and so on. The program allows the user to vary the distributions for arrival and service times, to have multiple service points, storage, different customer types, pre-emption rules, reneging, and priority rules. The program is designed to produce numeric and costed statistics at the end of the simulation process.

ACTION

- **use queuing theory ideas to establish parameters such as average time in the system, etc.**
- **examine the service design to check the way internal and public queuing systems operate in practice.**
- **are there ways of redesigning the process to optimise average times?**

SUMMARY

Queues and schedules are an important part of the customers' service experience. They are not inevitable or uncontrollable. In fact various choices can be made. It is possible by good design of the service to reduce queuing time significantly. If the situation allows the manager to move away from the first-come-first-served discipline, it is possible to cut down average queue lengths, average time spent in the system and so on by an important amount.

CHAPTER 9

Computers: the Service Industry Technology

'For the holders of conventional wisdom, innovation is not novelty but annihilation.'
Marshall McLuhan

Many specialised tools and machines are used by service organizations. Dentists, doctors and hairdressers have a highly developed range of hardware, as do restaurateurs, tree surgeons and retailers. One machine however is common to all service industry — the computer. Not only is it common to all, but such is its enormous present and future impact on the way service businesses are run, that it deserves a chapter to itself. (This chapter is written to describe the current state of a fast-moving technology at the beginning of 1992. The further away from the time of writing, the more things are likely to have changed. Before buying anything, those changes should be checked out in the specialist magazines.)

Thirty years ago, service industry used very little technology. In the accountants' offices, in the sales departments of manufacturing plants, in the branches of banks, there were calculating machines (electro-mechanical at best), typewriters and pens. Most techniques and equipment would have been broadly familiar to Scrooge's clerk Bob Cratchit. As a result, although service companies took an increasing proportion of the workforce, because their efficiency was static and that of manufacturing industry dynamic, the sector's relative output actually fell (O'Hagan 1991). The development of a machine specifically designed to handle the service industry's basic commodity, information, introduced radical changes. The computer, and more especially the personal desktop computer plugged into a network, is changing the face of service industries in ways which we are only just beginning to understand. Nowadays service companies are in some ways more technology dependent than industry. Perhaps 70–80 per cent of all investment in information technology goes into service environments.

Computers in one form or another have been around since the Second World War. One of the very earliest, the ASCC which first saw light in 1944, had over 750,000 parts, and took twelve seconds to divide two numbers. (If a modern machine takes twelve seconds to do anything, one suspects a system failure!) The first commercial general purpose computers went on sale in the early 1950s, with the café company J. Lyons in the forefront of development. The first IBM computer was launched in 1953. At this time there were perhaps 100 computers in the whole world, half of which were based in the States. By 1960 there were perhaps 10,000 computers in the world and by 1970 100,000. By 1990 the number was in millions. Typical usages for these early computers were payroll, invoicing, stock control, sales accounting and costing. From the beginning the computer was the service environment tool *par excellence.*

This has of course changed the face of the service industry. In the financial world, electronic connections have made the world a 24-hour financial market, with prices bouncing from Tokyo to New York to London and round again. The methods of dealing have changed too, with all sales and purchases data now being handled electronically and not on paper. Even instructions on buying and selling are computerized, with automatic sell instructions being triggered by falls, thus, it is believed, contributing significantly to the world market collapse of October 1988.

In retailing, electronic point of sale (EPOS) equipment using bar codes records sales as they happen, thus enabling shops to update stocks and identify high selling products on a daily basis. In Clery's, sales of the Miss Selfridge range of young women's clothes are monitored directly in the British warehouse. In advanced environments these sales automatically trigger re-orders not only from the head office warehouse, but also right through to the original supplier. The customers of retailers from DIY shops to clothes shops are benefiting from these ideas. In the large clothes chains in Britain, so-called Quick Response techniques have reduced the lead time for new products from weeks to days, at the same time enabling stores to carry more stocks while reducing stock mark-down losses. BHS (British Home Stores) for instance is now looking for a two-day repeat order turn-round from suppliers of best-selling items: Woolworth's have reduced lead times on new children's clothes from 10 weeks to 10 days.

For the suppliers of these firms, this means a drastic change: even large manufacturers now have to reconcile themselves to handling orders of 300–400 items when a few years ago 3,000 was a small run. To stay in business at all, they now have to be part of the electronic chain connecting the choices made by the final consumer in the shop to the original manufacturer. Once the network is set up, these electronic chains are attractively quick, easy and cheap to operate.

This is part of a long trend in increasing communication speed. In the 1850s, a written communication took several weeks to get to America; by 1900, the time was measured in days; by 1950, it took only hours, and now, by fax, only a few minutes or even seconds is required. In the computer world tenfold improvements in performance take years not decades. In 1981 the IBM PC worked at a speed of 100,000 instructions per second; by 1985 the IBM AT worked at a million instructions per second (MIPS); now you can buy machines working at 10 MIPS. MIPS on desk-top machines are expected to continue to increase by an order of magnitude every five years (see for instance the industry survey in *Byte* January 1990, 246).

The opportunities for improved service presented by this amount of power on desk-top machines have hardly been properly explored. Obviously, the service designer and manager must become and remain sufficiently computer literate to know if, when and how these extremely powerful tools can be used. The purpose of this chapter is to provide the basic information necessary to anyone wishing to investigate the use of desk-top computers in their business.

Computers can be considered under three main headings: hardware, peripherals, software. Each of these can be further subdivided:

- *hardware* refers to the machine that actually does the computing. Inside this box there are three aspects the non-specialist needs to be aware of. These are the microprocessor (the chip), the memory (RAM and backup) and the storage.
- *peripherals* are the stick-on bits that enable the user to communicate with the machine and vice-versa. These include at least the keyboard, the mouse, the monitor and the printer, and possibly other devices. In most cases these can be specified and bought independently of the basic machine.
- *software* is the set of instructions that make the hardware useful. It is by far the most important aspect of the computing scene. Everything a computer does has to be done through the medium of a program, or software. Software can be divided into operating systems, which are housekeeping programs, and applications programs such as word-processing, spreadsheets, databases, games etc.

The modern business computer market is divided into two broad strands. There is the IBM/DOS group, which owes its origins to the IBM PC launched in 1981, and the Apple/Macintosh strand which started in 1977 with the Apple II, and was continued by the Macintosh in 1984. Though a powerful and attractive machine, particularly in terms of user-friendliness and graphic capability, Macintosh have less than 20 per cent of the market. The DOS group by no means consists of only IBM; in fact the majority of machines sold in this group are not IBM. Hardware systems, peripherals and software are markedly different between the two groups of machines.

Hardware

Computers work by shuffling electronic 'messages' from place to place inside the machine under detailed instruction and at extraordinary speed. Each message is stimulated by the inputs from the keyboard or other source, converted into binary code, connected with other lumps of code, stored temporarily, reconnected, reconverted and displayed.

Computers gain their utility and power from their extraordinary speed of operation. For those (most of us) who cannot imagine what a million is, the modern desk-top machine's ability to process data at 12 million steps per second or more is completely incomprehensible. Yet this is the source of the machine's usefulness. Enormous speed enables the computer to manipulate the banal 1's and 0's of its basic arithmetic into calculations that provide real usefulness. Sheer repetitive speed turns the machine from a toy into a tool.

Microprocessors: The core of the computer's operation is the microprocessor, or chip. Otherwise called the central processing unit, the microprocessor is the key to the power and speed of the machine as a whole. Every instruction in the programme is separately fetched from memory and executed, one by one, inside the CPU. The speed with which it can execute instructions, and the size of the parcels of data it can handle are therefore the limiting factors on the machine.

The desk-top microprocessor market is dominated by two firms, Intel and Motorola. Intel founded the industry in 1971 with the 4004 (chips are numbered not named, originally from the number of transistors they replaced); their chips, now called the 80x86 series, are used in the IBM/DOS group of computers. The Motorola chips are used in Apple Macintoshs; they come in the 680x0 series. At the time of writing the latest in these two series are the 80486 and the 68040; each has the equivalent of a million transistors on a single chip. Chips with four million transistors are due in two years or so, and 16 million expected by the mid-1990s.

The higher the number of the microprocessor, the more advanced the machine. Generally this means that it goes faster, which in turn means that it can handle many more sophisticated instructions in the time you can be bothered to wait. Machine speeds are measured in MegaHerz (MHz); the higher the better. Current high speed machines work at 33 MHz. This speed is no more essential than the speeds you can get out of a Lamborghini; perfectly good work can be done on much slower machines. Indeed, some programs are written to run at a certain speed, and will not operate at any faster pace. It is common nowadays for machines to have the option of two speeds to accommodate this.

Chips are also compared by the size of the packet of data they can handle. This is measured in bits: eight bits is equivalent to one letter. Clearly a machine which can operate in 32 bit units is considerably stronger than an older 16 bit machine.

Memory: After the microprocessor, the next key variable is the size of the Random Access Memory, or RAM. Memory is used by the computer to store the results of the processing activity in the CPU. Desk-top computers have two forms of available memory: the RAM, which is inside the machine, and backup storage (hard or floppy disks). RAM is fast but volatile: switch the machine off and everything in RAM is instantly wiped out. Backup storage is slow but permanent. Programs have to be reloaded into RAM from backup storage every time the machine is turned on.

Some key dates in computer history

1842 Charles Babbage publishes his sketch of the Analytic Engine.
1848 George Boole, Professor of Mathematics in Cork, publishes his first book on Boolean algebra.
1937 British mathematician Alan Turing publishes the concept of the 'universal machine'.
1940s first generation computers: ASCC, ENIAC, EDSAC etc.
1950s second generation: large commercial machines based on transistors.
1960s third generation: based on integrated circuits, such as the IBM 360 series.
1971 first microprocessor chip, the Intel 4004, launched.
1975 the Altair 8800, the first microcomputer launched as a DIY kit.
1977 Apple II, with a maximum of 48 K RAM launched; CP/M, the first widely popular operating system launched.
1978 5¼" disk drives introduced.
1979 VisiCalc, the first spreadsheet program, launched. This program changed microcomputing from a hobby to a business concern.
1980 Hard disks, called Winchester drives, launched.
1981 IBM PC launched, expandable to 64 K RAM, and a single floppy drive; company expects to sell 250,000 over five years — however there were often months when that number were sold.
1982 First IBM clones appear.
1983 Apple introduce the mouse to microcomputing.
1984 Apple Macintosh launched.
1985 Toshiba 1100 laptop.
1987 Lotus 1-2-3 announced.
1988 Havoc caused to 6,000 US computer sites by virus introduced into ARPANET network by 'bored' student.
1989 Much upgraded Apple Mac IIcx launched with great success.
1990 Microsoft Windows 3.0 finally shipped.
1991 IBM and Apple announce future cooperation.

The CPU requires constant access to as much RAM as possible; the larger the packet size it's handling, the more memory it requires. RAM can also be used to store parts of the program. This enables the CPU to get its instructions direct and store results internally rather than having to go to the disk. The difference between the two is like the difference between remembering where to go and having to look it up on a map.

RAM is measured in thousands or even millions of bytes; the more you have, the better. This chapter is approximately 6,000 words, or 40,000 bytes; like many modern programs, the word-processing program requires a minimum of 512,000 bytes of RAM to run efficiently. This is written 512 K. 16 bit microprocessors can handle up to 1 megabyte (MB) of RAM (1,000 K or 1 million bytes); 32 bit machines can handle considerably more. This enables the computer to access several programs simultaneously, flashing from one to the other.

Backup or permanent storage comes in two forms, hard disks and floppy disks. Hard disks are usually installed in the machine on delivery. They are measured in terms of their storage capacity: typically 20, 30, 40, 60 MB. Hard disks are not as fast as RAM, but they are permanent (the data does not disappear when you switch off, as does RAM), and considerably faster than floppy disks. The critical performance factor in a hard disk is how long it takes the machine to access a particular piece of data. This access time is measured in thousandths of a second (ms), the less the better; anything greater than 65 ms will be irritating for referring for instance to a large database.

Hard disks do not however last for ever. Sooner or later operator error, mechanical failure, voltage variations, heat or physical damage are likely to cause the disk to crash, taking all your programs and data into sudden and unexpected oblivion. As a safety measure, all programs and all data should be copied regularly. This can be done in various ways: in small personal systems it may be sensible to use the hard disk for programs only, keeping all data on floppies (with copies). If this is not possible, then tape-based copies of the data will have to be taken. How often you do this depends on your estimate of the risk and how much it would cost to reconstruct the data. If you do experience a crash, do not press any keys at random. Whatever the computer is going to do on its own it will already have done before you notice there's anything wrong. Take a backup instantly. Specialist software such as Norton Utilities or PC Tools are available to help recover what is recoverable, which in some lucky cases may be virtually everything.

Floppy disks come in two forms, the 5¼" and the attractively robust 3½" which is becoming standard. The standard PC double sided/double density 5¼" floppy has a total capacity of 360 K of data (in practice about half this book). The single sided 3½" used in the Macintosh carries 400 K. Other formats with greater carrying capacity are available for both sizes. You will need a floppy disk drive to install new programs; afterwards the key advantage of floppy disks is their portability. Floppy disks also have a limited life; regular copies should be taken. As with hard disks, utility programs are available to help you to recover all or most of what was apparently lost in the event of damage.

Input/Output: The BIOS, or basic input/output system, is controlled by the instructions built into the ROM (read-only memory). This is the fundamental mechanism by which the computer reads from the keyboard, displays things on the screen, and generally handles the housekeeping. If the BIOS is not compatible with, say, broad IBM specifications, it will not be able to handle all the available software. The simplest way to test this is to install and run a program such as Lotus 1-2-3 on the machine; if Lotus runs, most other things will.

Peripherals

However fast the computer is, its impact on the real world comes through the screen, the keyboard and the printer. They therefore require special attention.

Printers: The slowest of these three is the printer. This will have to be bought separately from your computer, and there is a vast range of products to choose from. Two print technologies dominate the market: dot matrix and laser.

Dot matrix printing builds up each letter by a series of pinheads which impact the paper through an inked ribbon. The quality of the image depends on the number of pins used to make up the character matrix. For ordinary business use, 24 pin is the minimum acceptable. With 24 pins, the printer can produce various fonts, including OCR-B and a script. Most dot matrix printers can operate in two modes, draft and Near Letter Quality. NLQ is achieved simply by overprinting the characters twice using the same pins to get better resolution. NLQ is therefore considerably slower than draft quality. Advertised speeds of less than 180 characters per second in draft mode are unlikely to be satisfactory in office use.

Dot matrix printers are available in two standard carriage sizes: the normal 80 pica characters per line (on an A4 sheet) and the extended 132 characters. The larger carriage is invaluable if you have to print out wide spreadsheets. Two forms of paper feed are available, sheet by sheet, which is elegant but slow, and tractor feed, using special paper. Before you commit yourself to a printer, check that it is compatible with your software. If the printer is capable of emulating the standard manufacturer's mode, Epson, it is likely to be widely usable.

Laser printing also builds up the image or letter by a series of dots, but in this case it works off a grid of 300 x 300 dots to the square inch. This gives a 'near typeset' quality. (This book was set on professional typesetting equipment using almost identical laser-based technology, but to a density of 1,200 dots per inch.) The lasers control the dots, which are then reproduced by similar technology as the photocopier. Laser printers are quiet, reasonably fast (depending on the density of the page: a page of straight text will take much less time to reproduce than an elaborate graphic image), and expensive. Their

CASE STUDY: ADVANCING THE QUALITY ETHIC AT THE ULSTER BANK

The Bank recognises the need to engineer a variety of organisational interventions at regular intervals to reinforce constantly the quality ethic. It has used a variety of means to date some of which (the customer panels) have already been mentioned. Other ideas and interventions have included:

- Introduction of Quality Action Awards in which any one individual may nominate any other individual or group of individuals for Awards. In the first two years of the Programme over 300 staff received substantial Awards.
- The annual Chief Executive's Cup for Quality Service awarded to the QSAT which has shown most endeavour and motivation in tackling a local quality issue.

  1989 Winners Wexford
  1990 (Joint) Donegal and Maghera.
- The setting up of Area Quality Councils in each of the Bank's Six Areas throughout Ireland chaired by the local Area Directors and involving them in formulating the overall direction of the Quality initiative.
- The production of a Quality Service Newsletter (just being introduced).
- The organisation of a Quality Service Visioning Workshop in November 1990 to identify the objectives for Quality Service over the next 2/3 years.
- Building the Quality message into regular letters from the Chief Executive and the Chairman.
- Occasional high profile events in relation to a variety of products such as the launch of Standards.
- By the use of video as a means of communication. The following is typical of the video stock produced by the Bank and available to staff:
  - Customer Survey by Paul Clark in O'Connell St Branch and Corn Market Branch
  - Quality speech delivered by Feargal Quinn to Senior Management
  - Chief Executive on Quality Service Standards
  - Getting it right first time Drogheda Branch Training Video
- Area Directors audit branches on Service Quality on regular visits
- The Bank Inspectors are now required to assess branches and departments in relation to their adherence to quality standards and to make constructive comment in this regard.

Source: Ulster Bank

price is coming down all the time, though, and they certainly do produce a very professional image if you are communicating with the public. Desk-top publishing is dependent on laser printing. There is no need for each computer to have its own printer; because the printer is likely to be in use for only a fraction of the processing time, it is highly practical to attach several machines to one printer, thus justifying the cost of a grander machine than might otherwise be sensible.

Keyboards: The next peripheral to consider is the keyboard. Normally this and the monitor will come with the machine you buy, so the question of choice hardly arises. It is always possible to buy a keyboard separately, and this may be worth considering. If you are going to enter a lot of numbers, a separate number pad (like a calculator laid out to the right of the keyboard) is helpful. Function keys (numbered F1 to F12) are used in many programs for special instructions: make sure you have enough, some keyboards have only 10, not 12. The feel of the keyboard, the click the keys make when you hit them, and the 'travel', or the depth they sink, are personal matters, but they should be explored.

Many modern programs offer the choice of mouse-driven input for a whole series of instructions. The mouse is a pointer with buttons on it. It is about the size of a cake of soap. When you run the mouse across the table an arrow on the screen echoes the movement; when you reach the instruction displayed on the screen you click the buttons and the instruction is activated. Although this is slower than typing instructions, many people prefer to use the mouse. More and more programs are likely to be run by this method, the Graphic User Interface (GUI), as it is called.

Monitors: The last important peripheral is the monitor. This comes with the basic machine, so is not usually a matter for decision. Once again, however, there are choices to be made, depending on the software you wish to run. The first choice is between colour and monochrome. This is not the same as the difference between colour and black and white television: for most uses a monochrome monitor is actually less glaring for the eyes. Monochrome may be green-on-black or orange-on-black; both of these can be reversed to black-on-green or orange. Some people believe that orange-on-black is the least demanding on the eyes.

Nowadays however, many new programs use the extra discrimination possible in colour and graphics to make their points. Although monochrome is generally fine today, the future is in colour. If you choose colour, or if you wish to use graphic-based programs on IBM/DOS machines, you will need a graphics card added to the electronics inside the box (done by your friendly consultant/supplier), or it may be included. These vary. At the lowest level a Hercules Graphics Adapter Card will enable you to display graphics but only in monochrome. Various more expensive cards will enable you to have graphics and colour. These start with the Colour Graphics Adapter (CGA) which gives you only a limited amount of graphic colour; the next level up is the Enhanced Graphics Adapter (EGA) card, which needs a special compatible monitor. This was the top of the range until the VGA card was launched. If you need colour, a VGA or Super VGA card, with the compatible monitor, is probably the best, though expensive, buy.

Software

A computer, like any other tool, is only of use for what you can do with it. If you do not have any nails, why do you want a hammer? The only purpose of a computer is to be able to run software that will enhance your profit in some way. Software, or programs (always spelt in the American way), are simply a list of instructions which tell the computer what to do. No instructions, no activity. The computer is a general purpose information handling tool. Software converts it from general to a specific purpose.

Software can either be specially written for your need or you can buy a ready-made package off the shelf. It is not normally feasible for small to medium sized businesses to have software written for them, though it is extremely practical to write small routines inside larger packages such as Excel or dBase IV. Even this is likely to require some professional help to get the best result.

In the early days of computing, off-the-shelf software was generally specific to a particular machine. This is the situation with the Apple Macintosh now. Every program run on a Mac has to be written or at least adapted for it. Machines in the much larger IBM/DOS group can use software designed for a whole range of machines, as long as certain basic requirements are met.

The business software market can be divided into several large fields:

- operating systems.
- word processing.
- spreadsheets.
- databases.
- other applications.

At the physical level, all computer activity consists of a series of on/off switches combined into logic gates, into arithmetic instructions and so on. Current either flows or it does not. Using an elaborate series of gates built into the microprocessor, arithmetic, logic, comparison, data movement and so on can be performed. These flows of current through these gates is dictated by the program, which at this lowest level is expressed as a series of 1's and 0's representing on and off. This is called machine code, and is clearly almost impossible for humans to work with. Much easier is the set of mnemonic codes (such as ADD and LOAD) called assembly language, which is translated inside the machine into machine code. Assembly code is however far from human, so most programs are written in a language such as BASIC, C, or PASCAL. They are then compiled, and/or translated into machine code using a special translation programme. Most computer users will never need to get anywhere near program writing, or if they do, the programs will be written inside other programs such as a spreadsheet program or a database program.

Operating systems: Early in the computer boom programmers were spending an inordinate amount of their time writing routines to perform a variety of mundane housekeeping and data moving tasks. Everyone who wrote a program for a particular chip had in effect to re-invent the wheel. These routines were more to do with the chip than the application details of the program, so 'operating systems' that covered all these standard, common, tasks were written for specific chips.

The best-known operating system (since IBM used it for the PC) is Microsoft DOS. DOS stands for Disk Operating System. This program is the key to the mutual compatibility of the IBM/DOS group of machines. The latest DOS is version 5; nothing before version 2.11 should be used. Some programs will not run unless the DOS is 3.x or greater.

The Apple Macintosh machine family have a much more comprehensive specification. Although Apple do not write software themselves, they have gone to great lengths to ensure that every program written for the Mac has an identical look and feel, based on a common use of screen icons (little pictures of actions and instructions), heavy use of the mouse, pull-down menus etc. even down to small conventions of presentation. This has the huge advantage that when you have learned one Mac program, you are a good way to learning all the others. The heavy and effective use of graphics, using the power of the Motorola 68000 chip, accounts for much of the popularity of the Mac, especially when it is combined with desk-top publishing programs.

Microsoft have recently moved into this same area, with their operating system Windows. This adopts many of the same icon/mouse/pull-down menu systems as the Mac. The other key operating system is OS/2, now in version 2.0. These two programs are full of features, but are suitable only for the larger desktops. OS/2 2.0 requires 10 to 15 megabytes of hard disk, and at least 4 Mb of RAM. As much as three quarters of the program can be taken up with presenting the information to the user, and only a quarter in actual number crunching.

Operating systems are the key to effective computer use. Because they are concerned with background activities, with file handling and sorting, etc. many people learn a few simple DOS commands and ignore the program thereafter. This is a mistake. It is particularly a mistake when as a result of expanded memory and larger chips, operating systems are providing more and more facilities. Time spent understanding your operating system greatly increases the use you can get out of a machine. Not to understand the operating system makes you as helpless as a driver who does not know anything about the car engine except where to put the petrol.

The dreaded viruses lurk in the operating system. Virus is a generic term for programs written by clever and destructive programmers. By interchange of disks they spread from computer to computer. They are designed to lie unseen in the operating system, and be activated at specific prompts. Some are harmless: they might, for instance, take five minutes to draw a large and silly picture on your screen every 1 April. Others are very destructive: a famous one instructed the computer to reformat the hard disk, thus wiping out all programs and data, every Friday 13th. (To defeat this, users started to change the date on the internal clock to Thursday 12th; a new virus was quickly written which overcame this evasive tactic by making the wipe-out occur ten days after Friday 13th.) Others slowly block up internal memory with uselessly replicated instructions.

Like human viruses, you can only get computer viruses by certain types of behaviour. If you never put anyone else's disk into your computer or network, you will not get a virus. Unfortunately this will severely limit the utility of the machine. Computing after all is about exchanging information. The worst (innocent) carriers are computer engineers and consultants, who take test disks and the like from machine to machine. Games players are the second most common source. Specialist programs have been written to search out and neutralize the several hundred known viruses.

Word processing: The most popular use for computers is word processing. Over 40 per cent of machines are used for nothing else. This is basically using the computer as if it were a sophisticated typewriter. However because of the computer, word processing programs can do a number of things that traditional typewriters cannot. For instance all word processing programs have a block move facility, whereby a paragraph can be moved from one place in a document to another without retyping.

This can be particularly useful in redrafting contracts with standard paragraphs. Another extremely useful function is search and replace: if you write a proposal to company A, and then wish to use the same proposal for company B, the search and replace function will automatically replace the 'A' references with 'B' references.

Other functions commonly included in a word-processing package include page layout, automatic numbering (and renumbering) of paragraphs, office functions such as forms and merging a series of names and addresses into a standard letter and automatic spelling checks. Many word-processing programs incorporate more and more desk-top publishing or advanced presentational features in their design. These often claim to be able to show on the screen exactly what will be printed out; this claim should be treated with caution. Typical word-processing programs include Microsoft Word, Wordstar, Wordperfect.

Spreadsheet: It was a spreadsheet program, VisiCalc, that changed microcomputing from a specialist's toy to a management tool. Business people, vague about the relationship between hardware and software knew they had to have VisiCalc; if they had to buy a computer to make it work, so be it.

A spreadsheet is fundamentally an enormous grid (like a chess board).

HOW A SPREADSHEET PROGRAM WORKS: CALCULATING A QUALITY INDEX

	A	B	C	D	E	F	G
1			Defectives found by inspection				
2	Date*	Index value	No of Critical*	No of Major*	No inspected*	Critical %	Major %
3	Jan–91	96	14	91	8599	0.16%	1.06%
4	Feb–91	96	8	115	9477	0.08%	1.21%
5	Mar–91	98	7	114	18942	0.04%	0.60%
6	Apr–91	98	3	119	16400	0.02%	0.73%
7	May–91	96	5	119	8814	0.06%	1.35%
8	Jun–91	95	14	59	5711	0.25%	1.03%
9	Jul–91	97	5	66	8169	0.06%	0.81%
10	Aug–91	98	14	48	11199	0.13%	0.43%
11	Sep–91	97	7	57	7874	0.09%	0.72%
12	Oct–91	96	2	49	3882	0.05%	1.26%
13	Nov–91	99	4	11	6439	0.06%	0.17%
14	Dec–91	97	1	33	4159	0.02%	0.79%
15	Jan–92	97	22	67	13487	0.16%	0.50%
16	Feb–92	96	12	74	8268	0.15%	0.90%
17							
18	* The columns marked with a * are input manually; the rest are calculated by formula						

Quality Index Chart drawn automatically by the program

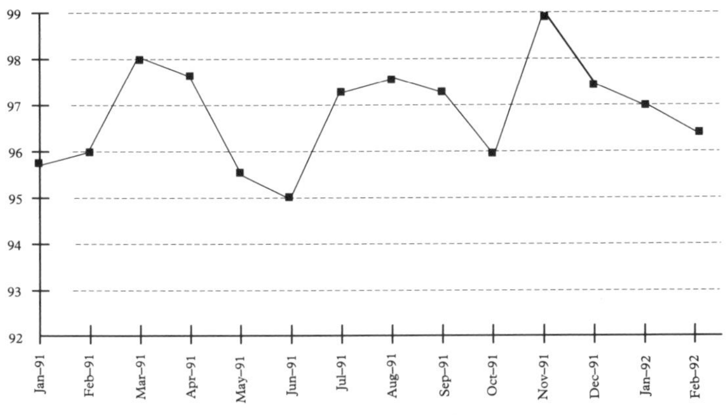

The Formulae in the Cells

	A	B	C	D	E	F	G
1							
2							
3	33239	=100–SQRT ((((1000*F3*50)^2)+((1000*G3*10)^2))/1000)	14	91	8599	=C3/E3	=D3/E3
4	33270	=100–SQRT ((((1000*F4*50)^2)+((1000*G4*10)^2))/1000)	8	115	9477	=C4/E4	=D4/E4
5	33298	=100–SQRT ((((1000*F5*50)^2)+((1000*G5*10)^2))/1000)	7	114	18942	=C5/E5	=D5/E5
6	33329	=100–SQRT ((((1000*F6*50)^2)+((1000*G6*10)^2))/1000)	3	119	16400	=C6/E6	=D6/E6

Notes: A = dates expressed in numerical form.
B = the index formula weights critical defectives at 50 and major at 10.
C, D, E = data input columns.
F, G = critical and major defectives as % of total inspected.

Each of the squares can contain either text, a number or a formula. If it is a formula, then the value depends on the number(s) in other squares. Thus if square A1 had the number 5 in it, A2, 6 and A3 the formula =SUM(A1+A2)*3, then the machine would calculate 33 and insert it automatically into A3. If A1 was to change, say to 6, then A3 would change instantly to 36.

This simple idea is extremely powerful. With it models of all sorts of business situations can be made. The effects of all sorts of changes can be quickly followed. A simple profit and loss account model for instance will allow the business manager to try out 'what-ifs': what if raw material costs went up? What if I raised prices by 3 per cent? What if wages went up, raw materials down, prices up slightly, unit sales down, and transport costs up, all at the same time? The cumulative effect of all these changes on the bottom line appears in seconds.

Modern spreadsheet programs such as Excel and Lotus 1-2-3 come with sophisticated graphic facilities, which enable charts and diagrams to be created both for display purposes, and for data exploration. For instance it is possible in a few seconds to graph your sales by week against other variables such as overtime, or weather patterns. Some of these ideas will work, most not; seeing it as a graph reveals the 'shape' of the data as nothing else will.

Databases: The third great area for personal computing is the database. This is fundamentally no more than an organized set of data. The computer can be used to search and shuffle the data in any number of ways. This might be some quite obvious set of data, such as the list of houses on an estate agent's books. The number of rooms, age of property, price, special features such as gardens, nearness to airports or sea, nearness of shops, are entered for each house. If a customer comes in looking for a modern three bedroom house near the airport with a large garden for less than £120,000, the agent simply types in the conditions, and any or all houses that meet the criteria are immediately found and listed.

This basic structure can with ingenuity make good use of the various types of information flowing through a company's hands during the service transaction. One hotel in Hong Kong used its database to record from the passport the birthdays of all its international guests, and sent them a card on the appropriate day. With electronic tills capturing the basic data, it would be possible for a supermarket to identify the clusters of products bought by certain types of customers (e.g. high spenders against low spenders).

There is of course an understandable nervousness about the misuse of database information, particularly in respect of private information. Databases which hold information about members of the public may need to be officially registered.

Networks: The real power of the computer is unleashed when several machines are combined into a network, sharing software and data. Networks combine hardware (machines, cables) with software (communications protocols, etc.) in a peculiarly effective way. Data such as sales information flows instantly from shop to stock-room, and from stock-room to manufacturer in a form that requires no further translation or input effort. Unlike written sales reports, it is immediately usable. For many service companies this technology has enabled them to deliver a new standard of service.

There is a price to pay however. Networks are vulnerable to attack from outside 'hackers', and they are troublesome to set up and manage. Networks come in two sizes: Local Area Networks (LANs), which combine the machines in a single office or campus (say within 1 kilometre) and Wide Area Networks (WANs) which cover a country or more.

The connections between WANs span the world in such uses as e-mail, the electronic mail service, ARPANET, and many others are well established. Modern networking is based on the international standard OSI Reference Model. This divides the various physical, connection, transmission and data levels of the message into seven levels. Each of the seven levels 'enables' the layer above it to operate. As a result of these complexities, network establishment and operation is a specialist task.

How to buy computers

1. Decide what you want the computer to do. Computers will not actually save time, but they will enable you to do well things that were not practical before. This includes such things as producing professional-quality client information leaflets, what-if models on spreadsheets, database stores of customer information to enable you to, for instance, select special interest customers for special promotions. They will not solve management problems, nor revive your business. It might be well to enlist the help of a consultant at this point, bearing in mind that many consultants have contracts with machine suppliers.

2. Find the software that will enable you to do what you want. Read the software reviews in the magazines such as *PC World*. Ask around. Check out the hardware and operating systems it will run under. Does it need special graphics cards, colour monitor, extra large memory? What support will you get from the supplier if you get into problems?

3. Find a machine that will run your software. The answers to the questions above will give you a clue as to the kind of machine needed. Compatibility, the ability to run software and, therefore, data on two different machines, requires:

- DOS compatibility.
- BIOS compatibility.
- microprocessor compatibility.
- program compatibility.
- ideally, floppy disk compatibility (i.e. size, density).

(This last condition can be waived if a direct cable link can be set up.)

4. Find out who can supply the computer you've chosen. You should have access to some kind of support both for hardware and software. Nothing is more infuriating than an unresponsive computer, especially if your valuable data on disk can only be accessed through this machine.

ACTION

- **computers are too important to leave to specialists: subscribe to a computer magazine (*PCW*, *PC World* etc.) and see how they can change your business.**
- **specialist software is available for many service industries: is yours one?**
- **how could a computer improve the delivery of service to your clients/ customers?**

SUMMARY

The computer is revolutionising the way service industries are being run, from legal practices to hairdressers. The impact is particularly strong in retailing. It is therefore imperative that service managers and service designers are aware if, how and when computer power can affect their business. This chapter introduces the basic facts about computer hardware, software and peripherals.

CHAPTER 10

The Service Delivery Process

Between the idea
And the reality
Between the motion
And the act
Falls the Shadow . . . this is the way the world ends
Not with a bang but a whimper.
T.S. Eliot *The Hollow Men*

As we have seen in Chapter 2, the service delivery process is governed by the service brief. Three statements (which, to conform to the guidelines of ISO 9004, must actually be written and controlled documents) lead from the service brief. These are:

- the service specification, which was discussed in Chapter 7.
- the service delivery specification.
- the quality control specification.

The last two will be discussed in this chapter.

The service brief identifies the customers' needs. The service specification details the activities that must be undertaken to meet those needs, and the service delivery specification explains exactly how the activities are to be carried out. The quality control specification lays down the checks and controls necessary to ensure consistency.

There is one further requirement to which thought must be given. Clearly there is no use in simply creating these specifications and documents without building in some method of long-term validation. This is in line with the general quality assurance approach. It is not enough simply to do something, you must also check the result. The service no doubt was fine when it was first devised, but does it still meet the customers' requirements? Are all recent codes of practice and legal requirements satisfied? Are the resources still adequate to meet demand? 'Validation', says the Guideline (6.2.7), 'should be a planned and documented activity, and should include considerations of actual field experience, impact of modifications in the service and processes, impact of personnel changes, adequacy of procedures, instructions, guides and proposed modifications.'

This process will very likely suggest some modifications to the existing practices. At this point one of the most difficult culture changes implied by Total Service Quality comes into play. In an undocumented system, a manager can make changes quickly and easily; any change, whether suggested by inspiration or circumstances, could be tested and imposed very quickly. In a documented system, every change must be properly recorded and published. This can be cumbersome. On the other hand, with formal recording and promulgation of the change, a manager can be sure that everyone is aware of it. There is no excuse: 'nobody told me' does not work any more. The first key is to maintain one-to-one coincidence between what is written in the procedures and the actual practice. This can be done only if people are fully conscious of the written system. If this line is not held, the total service quality system will begin to collapse.

Service delivery specification

The ISO Guidelines (6.2.4.1) describe the service delivery specification as follows:

'The service delivery specification should contain service delivery procedures describing the methods to be used in the service delivery process, including

- a clear description of the service delivery characteristics that directly affect service performance.
- a standard of acceptability for each service delivery characteristic; (These two are the link with the service specification.)
- resource requirements detailing the type and quantity of equipment and facilities necessary to fulfil the service specification.
- number and skill of personnel required.
- reliance on subcontractors for purchased products and services.

The service delivery specification should take account of the aims, policies and capabilities of the service organization, as well as any health, safety, environment or other legal requirements.'

The Guidelines envisage this specification to be a natural outcome of the service design process. 'Design of the service delivery process may usefully be achieved by subdividing the process into separate work phases supported by procedures describing the activities involved at each stage. Particular attention should be given to the interfaces between separate work phases. Examples of work phases involved in services are:

- providing information about services offered to customers.
- taking the order.
- establishing provisions for the service and delivering the service.
- billing and collecting charges for the service.' (6.2.4.2)

For each of these phases there must be a written procedure, probably based on a flow chart, which itself is derived from the overall service blueprint. The procedure should answer the following questions:

- what resources are required to perform the phase (staff, skills, equipment)?
- what inputs (quantity and quality) are required from internal and external suppliers?
- what internal or external customer need(s) does this phase relate to?
- how will we know if the need(s) are met?
- what operating procedure will ensure that the need(s) are met?
- what health, environmental, safety or other legal requirements must be addressed in meeting the need(s)?
- what connections must be made with internal and external customers to complete the task?
- what records show its completion?

Any procedure that answers these eight questions will meet the requirements of the ISO standard.

Quality control specification

In the days before calculators and desk-top computers, maths teachers were fond of identifying two steps in arithmetic: doing the sum, and checking the answer. A sum is not complete until you have proved to yourself that the answer is right. One favourite method of checking was the digit sum check. If for instance you had to multiply 37 by 4,287, you should arrive at 158,619 as the answer. To check it by the digit sum method, add the digits of 37 together repeatedly until you come to a single figure (3 + 7 = 10, 1 + 0 = 1) and the digits of 4,287 (= 3); multiply the results (1 x 3 = 3). Then find the digit sum of your answer: (1 + 5 + 8 + 6 + 1 + 9 = 30 = 3). If the answers are the same, your calculation is correct. Not only do you have the right answer, but you can be sure that you have (see Trachtenberg 1968).

This is the contribution that quality control makes to the service delivery procedure.

The ISO Guidelines (6.2.5) state that 'quality control should be designed as an integral part of the service processes: marketing, design and service delivery. The specification developed for quality control should enable the effective control of each service process to ensure that the service consistently satisfies the service specification and the customer. The design of quality control involves:

- identifying the key activities in each process which have a significant influence on the specified service.
- analyzing the key activities to select those characteristics whose measurement and control will ensure service quality.
- defining methods for evaluating the selected characteristics.
- establishing the means to influence or control the characteristics within specified limits.'

The Guidelines go on to give an example showing how key activities, characteristics and evaluating methods are connected: 'The application of quality control principles is illustrated in the restaurant service example below:

a) A key activity to be identified in a restaurant service would be the preparation of a meal and its effect on the timeliness of the meal being served to a customer.
b) A characteristic of the activity requiring measurement would be the time taken to prepare the ingredients for a meal.
c) A method of evaluating the characteristic would be sample checks of the time taken to prepare and serve a meal.
d) The effective deployment of staff and materials would ensure that the service characteristic of timeliness was maintained within its specified limits.'

In designing the restaurant service the management identifies tolerable and intolerable waiting times. Clearly these would be different for a fast food place than for a top hotel. The management decide to measure the time between giving the order and the service of the order; how long the ordered meals take to prepare, in fact. To measure this time, they take samples of meals during the day, and plot this time on a chart. This enables them to notice how time taken expands and contracts over the day (longer in very busy times and so on). As a result they are able to plan the employment and use of staff to ensure that the timeliness targets are met.

Ideally each procedure should have some check or control built into it. In Japan this principle is called *poka-yoke.* This translates as foolproofing. The theory is that the design of the job should as far as possible have built into it methods and operations that make it harder to do the job wrong than right.

For instance a manufacturer of board games may want to check that every game has the full set of components, so an automatic scales is used at the end of the assembly line to compare the weight of each box with a target. If the weight falls above or below the target tolerance, a buzzer sounds, the line stops, and the offending box is examined. A printer may wish to ensure that no blank pages are included in books. A scanner is installed that seeks a printed image on each page, and stops the machine if the image is not found. Result, no blanks. The inventor of the *poka-yoke* theory, Shigeo Shingo, has pointed out that one of the great advantages of *poka-yoke* devices is that they stop defective products from progressing to the next process. No value is added to already non-conforming goods. This brings a physical reality to the internal customer concept.

The *poka-yoke* system builds the check into the design of the job. The check is no longer a second stage (as in our arithmetic example), but integral to the design of the job. Parts to be assembled for instance are designed so that they simply cannot be assembled other than correctly; for instance, contacts are placed at the bottom of screw holes to ensure that all screws are run home.

The theory of *poka-yoke* was devised for industrial manufacturing environments. However the principle behind the system is equally valid for service environments. In one simple assembly job, Shingo found that workers were inclined to leave out one of a set of screws, which they picked one by one from a common box. He simply redesigned the operating procedure so that the workers picked all the screws for each assembly at once. It became immediately obvious if one had been left out. This kind of job design can build checks and controls into the simplest job. In the restaurant example above, for instance, it would be easy to arrange that every order coming into the kitchen is time-stamped as it comes in and when it goes out. This would then provide a check for every order, not just a sample. Orders that were running into time trouble would be clearly visible. If the recording was done on a computer, orders that were becoming late could be flagged for urgent attention. It may not always be possible to devise such controls; in that case a simple checklist of actions may be useful.

Whatever quality control devices are adopted, their prime function remains clear. This is to detect non-conformances as quickly as possible. There is also a secondary function. This is to establish quality records as the basis for analysis and improvement of the service. Every organization needs to have a self-conscious strategy for improvement at the heart of which is the gathering and analyzing of operation and quality data.

The five gaps between aspiration and performance

Between the creation of a proper service delivery specification and the actual moment of truth experienced by the customers there too often falls a shadow. Like a general sending troops into battle, the service manager must ensure that as well as a strategic plan, and tactical training, all logistic support is available, and that forces are concentrated where they will have the most effect. But at the point of service delivery, the moment of truth, contradictions arise between the service the customers expect, and the service the managers wish to supply.

These contradictions arise from the whole process of service delivery. They start with the management's perception of customers' needs, and end with the relationship between what the customer gets and what he or she wanted.

THE FIVE GAPS BETWEEN GOOD AND GREAT SERVICE

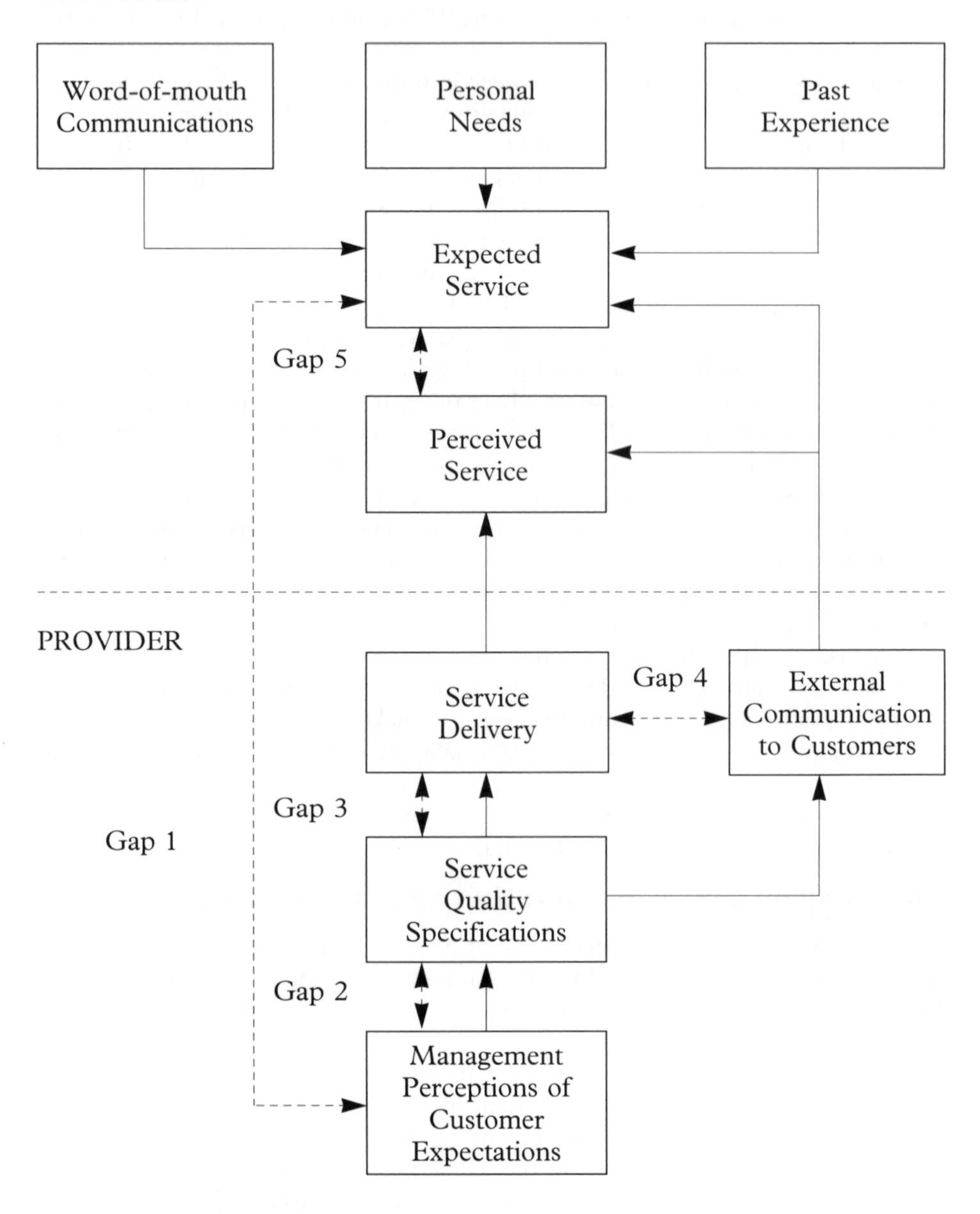

MODEL OF SERVICE QUALITY

Source: Zeithaml, Parasuraman, Berry (1990)

Five major causes of service quality shortfalls have been identified (Zeithaml, Parasuraman and Berry 1990). These are called the Five Gaps. They are modelled in the diagram, and are as follows:

- Gap 1 between customers' expectations and management's understanding.
- Gap 2 between management's perception of customers' expectations and service quality specifications.
- Gap 3 between service quality specifications and service delivery.
- Gap 4 between service delivery and external communications to customers about service delivery.
- Gap 5 between customers' expectations and perceived service.

The Five Gaps between service aspiration and service reality are pointers to management oversight of the process. If nothing is falling through any of these gaps, then you have a perfect service system.

Gap 1: Not knowing what customers expect

Many service suppliers start to analyze their service design by thinking what *they* would want from such a service. In some ways this can be a successful strategy.

Unfortunately as soon as one starts a business, one's mind is affected by the inside knowledge of the trade. An aspiring bookseller may start by offering wide stocks of current affairs and history and a rapid ordering service: after a few years in the business he or she has forgotten what it is like to be a customer, and only remembers the occasions when a book took eight weeks to arrive from Edinburgh, or a customer never returned after ordering a £45 book, or the fact that 80 per cent of the buyers apparently want only Jeffrey Archer and Joan Collins. As soon as you step on to the other side of the Line of Interaction, your view of the world changes, and that's not all that changes. The customers' expectations change constantly too. As a result it is essential to refresh constantly your view of what the customers want.

The gap between customer expectation and management perception of that expectation can also be caused by problems with internal communications in the company. The frontline service providers may be very conscious of customers' requirements or gripes. There is often a problem, however, in transferring that understanding to management. This is frustrating for both sides; for the frontline workers, because they are constantly exposed to customer dissatisfaction as a result, and for management who probably genuinely want to get things right, but are not aware of how to do so. In this context, perhaps the most frustrating and infuriating remark for a manager to hear is: 'Oh, I could have told you that!'

If there is no mechanism for pushing the information painfully gained at the frontline to the managers who design the service, the knowledge is wasted.

Managers who spend all their time in the offices on the sixth floor, or who build in layers of procedures and staff between the line of interaction and themselves are likely to become increasingly remote from the real customers' expectations. Claus Møller of TMI International tells the story of a US bank whose motto was *The Customer is Always First*, but which allocated all the best spaces in the car park to the chairman and the senior executives.

The simplest market exploration technique is that of analysis of complaints, used by 85 per cent of top Irish companies. All complaints, great or large, should be logged regularly, and the logs or a summary of them distributed to key staff. They should be logged in such a way that points to a solution; for instance details such as product, time, area the complaint arose in, name and address of complainant (to enable personal response) and other details should be carefully noted for all complaints. The relative seriousness is often evaluated by cost or other criteria such as safety, width of public impact and so on.

As we will see in Chapter 12 on continuous improvement, there are two reactions to a complaint. The first is the immediate 'fix-it' reaction. This should be done with as much panache and grace as possible, to achieve what Tom Peters calls the 'remarkable recovery'.

The second reaction is the internal one, the use of the complaint as an input to strategic planning. As Zeithaml, Parasuraman and Berry put it, 'Complaints offer opportunities for managers and contact personnel to interact with customers, thereby learning detailed and rich information about products and services. Although a goods firm, Proctor and Gamble recognizes the strategic advantages of providing service in the form of an 800-number that customers can call with problems. Through the telephone interaction, customer service personnel ask specific questions, get to the heart of customers' problems, and learn more about the company's products.

'This information is summarized and given to management as input when planning product or marketing changes. Senior vice-presidents also spend three hours a week answering the phone, hearing customers first-hand, answering questions and addressing the problems that are raised. Describing the value of the experience, one executive said, "I can't tell you what I do differently as a result of answering the phone, but I can tell you that no decision is made quite the same way."'

If you rely on complaints alone for information, does the fact that you did not get any this month necessarily mean that you did a good job? The data from TARP research programmes would throw some doubt on that, since up to 90 per cent of those irritated by a service never bother to complain at all. Ideally therefore the results of one market exploration technique should supplement and cross-check those of another.

Complaints are a rich source of information, but they are essentially reactive. To be more pro-active a company must use research. This might involve one of the many specialized techniques employed by market research companies, or perhaps a more simple form of customer comment card. Techniques of exploring markets by use of research are discussed in more detail in Chapter 2. Understanding the market is a continuous job, like painting the Forth Bridge. It should be carried out by a wide range of techniques, from the relatively cheap and simple such as analysis of complaints, to the more sophisticated such as multi-dimensional market research programmes. In establishing a market exploration strategy it is always important to remember that every technique presents its own problems of interpretation.

Whatever market research techniques are used, it is essential that all the results be widely disseminated and analyzed. The information gained from research programmes should be analyzed in the light of shopfloor information and vice versa.

Ten ways to open customer access to your organisation

1. Talk to the customers face to face.
2. Organise focus groups.
3. Set up a customer survey; on the phone, by card, by post.
4. Have suggestion boxes and other quick feedback systems in place.
5. Find out where the problems are and resolve them. Then tell the customers you have done so.
6. Establish a customer newsletter.
7. Make customer service and complaint handling a key results area.
8. Respond rapidly to complaints with overkill.
9. Make it easy for the customers to get refunds, to exchange goods, to make complaints.
10. Evaluate managers' performance on the ability to access customer feedback.

(Based on Martin 1989)

Gap 2: Between management's perception of customers' expectations and service quality specifications

Understanding customers' needs and expectations is the essential first step in the creation of Total Service Quality. The next step is to put these understandings into the form of a company service specification or statement of intent. To do this management has to make a serious commitment to meeting the needs identified. In particular this means devoting adequate resources to the job. This commitment can be shown in various ways. Some of the issues are:

- are service oriented managers promoted and rewarded more than others?
- is there sufficient automation to meet needs consistently?
- are upper and middle managers obviously committed to service?
- are documented procedures taken seriously?
- does the company establish clear service goals?

- are resources and personnel available to deliver the quality of service required?

This gap is perhaps the most serious and most common obstacle to the creation of a quality service operation. Over and over again writers on quality insist that without serious commitment from the top, *fully backed by financial and personnel resources*, all the quality aspirations in the world are useless. In company after company managers and frontline service deliverers know what customers want, but for various reasons they are unable to provide it. Reasons include problems such as:

- a perception of impossibility (as in: 'zero defects? . . . forget it!').
- ill-designed premises.
- company policies and customs stressing inward-oriented goals.
- union rules.
- market conditions.
- lack of capital.
- lack of management incentive or drive to make the necessary changes.

For many executives the ability to meet known customer requirements is made impossible by a thick mesh of these constraints. This inadequate management commitment to service quality may be caused by various factors, typically:

- a fear of change from established policies and methods to untried new ways.
- an over-concentration on cost control and short-term profitability.
- a failure to see the customer as a life-time stream of revenue rather than a single quick hit.
- a perception that the customers' requirements are unreasonable.
- an unwillingness to commit resources to customer service.
- a lack of middle management commitment.
- a lack of understanding of quality management techniques.

This last can be a powerful reinforcer of the previous attitudes. If the methods of service design, of service specification and service delivery specification are not clearly understood, managers can retreat easily into negative attitudes. Managers need to be convinced that effective specifications can be written. Equally, if the techniques of setting quantified quality standards based on clear characteristics are not effectively used, quality standards and commitment can disappear.

Gap 3: The service performance gap

However well the management understands the customers' needs, and however thoroughly the service design relates to that understanding, if the employee is unable or unwilling to perform the service tasks at the level required, the gap will appear between service design and service performance.

It is easy to underestimate the complexity of the task at the point of interaction between the customer and the firm. One study of life assurance sales people has distinguished six different steps in the process of completing a sale. These are:

- contact initiation: 5 per cent of all acts.
- building rapport, establishing common ground: 5 per cent of all acts.
- clarifying needs, exchanging information: 40 per cent of all acts.
- persuading: 30 per cent of all acts.
- closing the sale: 15 per cent of all acts.
- follow-up, ending the discussion: 5 per cent of all acts.
  (Wilkie 1986)

With this complex set of tasks it is not surprising that sales people occasionally fail to deliver on all the needs. Organizations that are highly interactive, labour intensive and perform in multiple locations are especially vulnerable to Gap 3. 'Opportunities for mistakes and misunderstandings exist when service providers and customers interact: both customers and providers experience and respond to each other's mannerisms, attitudes, competences, moods and language. Greater variability is also more likely in labour-intensive services than when machines dominate service delivery. Bank customers who use human tellers experience far more service variability than those using automatic teller machines. Finally when service is produced in a chain of outlets, quality control is complicated because the organizational layers between senior management and frontline service providers hinders two-way communication and make it more difficult to assess individual employees' performance.' (Zeithaml, Parasuraman and Berry 1990)

This Gap may be caused by any of seven specific factors. These are:

- lack of knowledge, training or skill to do the job.
- wrong people in the job.
- inadequate or faulty equipment.
- role conflict/job design (particularly where multiple functions contradict each other).
- weak or inappropriate supervisory systems.
- lack of worker identification with the company.
- lack of encouragement/permission for workers to use initiative.

Research shows that the most damaging of these factors is bad job design leading to role conflict. This typically occurs when service staff are asked to sell as well. Thus if bank tellers are asked to provide normal cash services and, simultaneously, advice about other bank products, their ability to provide a quick service will be impaired. If a sales/service engineer is also asked to collect debts, he will tend to do one or the other well, but not both. Other forms of role conflict can occur when staff dealing with walk-in customers are also asked to meet quick telephone answering targets, or where check-out staff are also expected to find obscure products, collect boxes and confirm prices.

Gap 4: When promises do not match delivery

Advertisers, they say, sell the sizzle, not the sausage. Unfortunately when one takes the product home, the appetizing sizzle is often not quite reproduced. This gives rise to a sense of customer disappointment: the promise has not been matched by the delivery.

Over-promising is a significant cause of customer dissatisfaction. Advertising regulations forbid over-promising in technical matters, but in unmeasurable quantities such as 'friendliness', 'quality' and so on, any claim is arguable. Over-promising does not only occur in advertising: brochures and sales people are equally vulnerable to unsupportable overstatements which eventually induce a sense of cynicism in the customer. Oddly enough under-promising or under-selling can also create a problem. Every company has a more or less elaborate set of checks and controls by which it ensures that the product is delivered. These are normally hidden from the customers, so that the company is giving an inadequate sense to the customer of the service provided. Many financial institutions claim to be friendly — customers would prefer them simply to be efficient.

USING THE TELEPHONE

In most cases the first contact a customer has with the company is through the telephone. That first impression is extremely important in establishing the company's image in the potential customer's mind. It is therefore important to establish good telephone style.

1. Get good equipment.
2. Establish the telephonist's job properly. If she has to type and act as receptionist, the chances are she'll be able to do none of the jobs to full satisfaction.
3. Train and establish discipline.
4. Set up operational norms, for instance:

- answer all calls at three or four rings, neither sooner nor later.
- smile as you answer; sound friendly.
- identify the firm and yourself.
- obtain the caller's name and company (*before* you see if Mr So-and-so is in).
- if it's a caller you do not recognise, check the name back quickly to ensure it is right.
- establish who the caller wishes to speak to.
- make the connection.
- if the connection is taking time, keep the caller informed every 20/ 30 seconds (don't disappear).
- if the connection cannot be made, take a message.
- use the caller's name in conversation.
- learn to recognise regular customers' voices.
- don't struggle with a bad line, close the call and ring back immediately.

Another cause of problems in this area is where communications inside the firm fall down. Typically this may occur when the advertising and marketing people start to promote a new service before the frontline people have been trained in how to deliver it. Customers come in through the door only to be met with vague replies, such as: 'Yes I saw that in the paper myself, but we have not been told about it yet.' In large multi-chain firms, part of the customers' expectation is derived from experiences in other units in the same chain. If I have had superb service at one hotel in a chain, I will be keen to go to another. If the superb service was not a group effort but stimulated by the local manager, the second hotel will let me down. I will feel that the chain as a whole, by failing to meet its own highest standards, has been over-promising.

Gap 5: Between customers' expectations and perceived service

The fifth gap summarizes the effects of the other four in the mind of the customer. In the model we can see how the other gaps, which are in the hands of the management to close, all affect this one. The management's objective must be to bring the customers' expected and desired service as close as possible to what is actually delivered. Understanding what is wanted, having the management will to achieve that, ensuring that the service specification is carried out and controlling what is promised are the keys to closing that final gap.

ACTION

- **are the key service characteristics covered by well-understood written procedures?**
- **can you establish foolproofing devices at these points?**
- **identify the working of the five gaps in your organisation.**

SUMMARY

The service delivery specification details exactly how the service is to be provided. It follows logically from the service specification, which details the sequence of service to be provided. One identifies what is to be done, the other how the various tasks are to be carried out. Allied to the procedures and operating instructions of the service delivery specification are the quality control specifications, which detail how the various checks and controls operate.

Between the customers' ideal and the actual service fall various gaps. Five major gaps are identified as causing service problems, each with characteristic causes and solutions.

CASE STUDY: STARTING A QUALITY SYSTEM

The transition from theory to practice is a complex undertaking in itself requiring vision, excellent project management, high calibre staff and well thoughtout communications. But it all starts with 'Commitment from the Top': without that drive and constant reinforcement from the General Manager, a Quality Programme is unlikely to succeed.

At American Express, such support and commitment has enabled our customers to benefit from a range of service improvements and marketplace reputation to be enhanced, as well as adding to the bottom line.

To begin with, measuring Service Quality is a five step development process:

1. Define those services provided as end products visible to our customers and classify them into service elements.
2. Develop measures of performances for each service element, in terms of timeliness, accuracy and responsiveness.
3. Determine an appropriate level of service, that is, what should the standard be for each measure?
4. Develop a universal measurement system for data gathering and reporting.
5. Lastly and most importantly, the Q.A. methodology identifies improvement opportunities.

Performance standards, or goals, are defined for all service measures taking the following factors into consideration.

- Customer Expectation.
- Competition.
- Legislation.
- Economics.
- Processing Capabilities.

Initially in 1987 we set standards on 20 key measures based on the findings above. Gradually we have taken on extra measures where necessary and have also reduced some standards.

On a monthly basis, we have devised a reporting system both local and worldwide. If we are off target, then we focus on corrective actions. If we have met standards then we ask 'Is there a better way?'

An example, let us use the process of approving an application for an American Express Card. Procedurally, this involved handling the application form a number of times in a number of departments. However, under the Quality Assurance approach, we defined the service from the Customer's viewpoint with the following measure:

'Percentage of application forms approved within 'X' days of receipt'. When we first saw the results of the measure, we realised that improvements were necessary not only to deliver the service quality but also to improve the bottomline. Revenue cannot be generated until the Card is in the Cardmember's possession! Within a few months of installation of the Quality Assurance Programme, the service on this measure had improved by over 50%.

Source: American Express

CHAPTER 11

Quality Assurance Systems

'Everything should be made as simple as possible, but not simpler.'
Albert Einstein

Quality assurance systems are the shell inside which quality grows. They provide the structure, the stiffening and the form inside which the company's aspirations to quality performance are made and strengthened. As with so many other shells in natural history, from time to time an organization has to discard one shell for a larger and more ample one, by revising and rewriting the systems. A full quality assurance system is in fact a guide to the operational procedures of the whole company. In this context the word quality does not mean 'good', or 'best' etc.; it means the operational ability of the company to meet the customers' needs. Inspection and QC are therefore only a small part of the quality system.

Modern quality assurance systems are usually based on the ISO 9000 series, which consists of the five main elements ISO 9000 to ISO 9004, and a number of backup documents, such as ISO 8402 which provides definitions for the often rather special use of terms in the quality vocabulary. The ISO series, as an international standard, is now the basis for most national quality management system standards such as BS 5750 in Britain, the standards in France, Germany, Italy and many other countries. The EC has adopted the standard as European standard EN 29000.

The basic premise of the ISO 9000 series is learnt from the scientific method. By carefully establishing working systems, by recording results and exploring improvements, a company can deliver consistent and improving standards of operational effectiveness. The key is in the written procedures. Properly handled, these are clear effective guidelines to all the key activities in the service process.

They identify exactly what should be done and when, and how it is to be checked. If things go wrong when once a good written system is in place, questions are reduced to three:

- was the system followed?
- if not, why not?
- was the system adequate to the circumstances (if not, change it).

The ISO 9000 series of standards consists of three standards and a number of Guidelines. The full set is broken down as follows:

- ISO 9000: for internal use, to enable management to decide which of the three models (ISO 9001–3) is best suited to their organization.
- ISO 9001: for assessing companies that design, make and service their products. Most service companies fall into this category.
- ISO 9002: for assessing companies that take an established design and produce it. This is the most common ISO registration category for manufacturing concerns.
- ISO 9003: for assessing companies that store and distribute products.
- ISO 9004: various guidelines and elaborations explaining how the basic standards 9001–3 are to be interpreted. 9004–2 relating to service industries is of particular interest.

ISO registration is always gained under one of the three basic standards ISO 9001–3. There is no special registration for service companies. The three basic standards are written as checklists for assessing a supplier company; they are not designed as self-examination or 'how-to' documents. This is more the role of the 9004 Guidelines. A new full set of all these documents should be purchased; it may also be helpful to acquire copies of the working drafts that ISO Technical Committee 176 is currently working on. These will, in more or less amended form, eventually become full parts of the system. Current drafts include guidelines for developing quality manuals. Information and copies should be sought from the National Standards Authority of Ireland (NSAI).

The Irish Quality Association's Quality Mark scheme is substantially based on the ISO series, though taking slightly different approaches on certain matters. This scheme is based on an annual audit, which looks for the same standard of excellence as the quarterly ISO audits, though not on such a formal checklist as the ISO series. In many ways the flexibility of the Quality Mark scheme is well suited to service companies.

In recent years two new schemes have been launched, mainly aimed at companies which already have established quality systems, and now seek higher targets. These are the Malcolm Baldrige Award scheme in the United States, and the European Quality Award Scheme. Both of these schemes assume basic quality management systems in place, and so they pay

CASE STUDY: TYPICAL CONTENTS OF A SERVICE QUALITY MANUAL

(1) Quality policy
(2) Amendments and distribution
(3) Management organisation and responsibilities
(4) Amendment Record Sheet
(5) Quality Assurance responsibilities
(6) Purchasing
(7) Product identification and traceability
(8) Product inspection and testing
(9) Instrument calibration
(10) Control of non-conforming product and material
(11) Corrective action
(12) Handling and storage
(13) Packing and storage
(14) Quality Records
(15) Training
(16) Document control
(17) Design control
(18) Contract review
(19) Management reviews and quality audits
(20) Hygiene
(21) Service standards
(22) Service specs
(23) Human relations
(24) Customer care/service
(25) Marketing/services

Source: Patrick Casey, Optician

significantly more attention to actual results achieved, rather than the systems approach of the ISO or Q-Mark auditors.

There are four key factors in a formal quality system:

Management responsibility: Management must document quality policy in relation to the service objectives of the company, the grade of service to be provided, the approach to be adopted in respect of service requirements and the role of personnel in the establishment of service quality.

Quality system structure: This is based on the Service Quality Loop. Service quality starts with a need identified by marketing. This generates a Service Brief which, as the Standard puts it 'defines the customer's needs and the related service organization's capabilities as a set of requirements and instructions that form the basis for the design of a service'. The service brief is then developed into a series of instructions and specifications for the service and its delivery.

These are two separate though connected functions. The service specification includes a complete working specification of the particular service, including a statement of the characteristics subject to customer evaluation, and a standard of acceptability. The service delivery specification, on the other hand, describes the procedures used in actually delivering the service. There may be several of these, each describing the procedures for a particular phase of the service delivery process. The Standard suggests for instance that procedures would be written for:

- providing information about services available to customers.
- taking orders; establishing provisions for the service.
- delivery of the service.
- billing and charging.

The above elements are the documented part of the service quality system.

Personnel and resources also spring directly from management resources. It is management's responsibility to ensure that sufficient and appropriate resources are made available to implement the quality system and achieve the quality objectives. The most important part of this responsibility is in the recruiting and training of staff.

Interface with customers is the key to the whole. This involves describing the service, its scope and availability; stating how much the service will cost; explaining the interrelationships between service delivery and cost; explaining to customers the effect of any problems, and how they will be resolved, should they arise; ensuring that customers are aware of the contribution they can make to service quality; providing adequate, readily accessible facilities for effective communication to and from the service organization; determining the relationship between the service on offer and the real needs of the customer.

How to create ISO standard documentation

The most unfamiliar and therefore troublesome part of establishing any formal quality system is the documentation. Many companies have perfectly good standards and operating practices, high and deserved reputations with their customers, and yet hardly any of this is documented. The problem is that the company depends completely on the training, knowledge and skill of its established staff. They may be very effective at what they do but unfortunately because systems are not clearly documented, it is often difficult to control change.

Documentation should be thought of as a precision control instrument for the delivery of a high and continuing level of quality. A skilled craftsman can estimate distances to a high degree of accuracy; a quality shop requires micrometers. Equally, a company like the one described above can frequently deliver good product; an ISO 9000 standard shop requires documentation to provide that extra control and precision.

A good quality documentation system contains four separate elements:

- the quality manual, a company-wide document, which summarizes how the overall quality system works.
- the quality plans or service brief, which describe how quality for each product or service is controlled, typically by way of a flow chart.
- the quality or service delivery procedures, which are written procedures and forms defining how operations, inspections and activities are to be carried out.
- quality records which record the results of specific inspection, training, etc.

These four elements form a hierarchy of control which ranges from the overview level of the manual, to the action-by-action detail of the procedures and the records. At the very top level the manual lays down the controlling ideas of the quality system. It can be compared to the National Constitution. Like the Constitution, it should be designed to stand above day-to-day changes in procedures.

Quality plans, the second level of control, describe the sequence of activities and controls for each product or service. Typically this will be presented in flow chart form, identifying Activity 1, check, Activity 2, check, Activity 3, check, and so on. These tend not to change much.

Procedures, which can be compared to laws, detail how each of these activities and checks are to be carried out. In a dynamic environment, these are subject to frequent improvement and change; as a result the rules for changing procedures and forms should be as light-footed as possible. If a department is inhibited from introducing improvements by cumbersome formal rules, the tail is wagging the dog. (Guidelines for document control are dealt with later in the chapter.)

The main difficulty in creating procedural instructions or standard forms lies in how much detail to include in procedures. The guideline is to create a formal procedure when, and only when, product or service quality will suffer if the full procedure is not carried out. For instance whether an operative uses the left or the right hand for a task is unlikely to matter; whether or not they are properly trained will.

The quality management standard ISO 9000 defines these four elements as follows:

Appropriate quality system documentation includes the following:

1. **Quality manual:** This document, or collection of documents, should provide an adequate description of the quality system as a permanent reference. Its main focus is to communicate the commitment to quality and to demonstrate effective compliance with required quality standards. It should contain:

- title, scope and field of application.
- table of contents.
- introductory pages about the company.
- manual (including revisions and issuing information, approvals, and a page identifying amendments).
- the organization's quality policy.
- the organization chart etc.
- quality plans.
- a description of the quality system, including all elements and provisions that form part of it (typically based on the chapters in ISO 9001–3).
- definitions, index, etc.

2. **Quality plan:** A quality plan details the specific sequence of activities and controls that a particular service or product will require. In particular the plan should identify:

- quality objectives for the service.
- management responsibility.
- how the various requirements of ISO 9001–3 are met.

3. **Quality procedures:** These are written procedures the absence of which would adversely affect the quality of service. They define how activities in the service organization are to be conducted. Before it is implemented a draft procedure should always be agreed by the personnel responsible for issue, those who will implement it and those who will interface with its operation. A typical procedure would include:

- document control details: document number, title, revision number, revision date, approval status.
- purpose and scope; why, what for, area covered.
- responsibility: who will implement the procedure.
- procedure: step-by-step details of what needs to be done; mention exceptions or specific areas of attention.

- documentation: documents or forms associated with the procedure.
- records: records generated, where kept and for how long.

4. Quality records/work instructions: Records and instructions derive from the procedures, where their use and retention is detailed. Work instructions may be derived directly from the 'procedure' section of the relevant procedure. Records provide information:

- on the degree of achievement of the quality objectives.
- on customer satisfaction and dissatisfaction with the service quality.
- about the results of the quality system for review and improvement of the service.
- for analysis to identify quality trends.
- for corrective action and its effectiveness.
- on appropriate sub-supplier's performance.
- on the skills and training of personnel.

The records should be:

- verified as valid before acceptance and storage.
- readily retrievable.
- retained for a designated period.
- protected from damage, loss and deterioration while in storage.

There is also a requirement in the standard that documentation be controlled.

Setting up the documents

Every company is different; consequently there is no standard way to approach the setting up of a quality system.

Occasionally the task is delegated to one manager who spends three months locked in a room and then triumphantly emerges with a completed document. This is likely to prove satisfactory to those companies who require a quality manual only because their customers demand it, not because they want one themselves. It is unlikely to work in companies with good internal motivation. Why should anyone pay attention to a document they had no hand in creating?

A consensus-based method of evolving the manual and procedures is much more likely to prove successful and lasting. There is an important reason for this. The critical relationship between the plans, procedures and 'real life' must be maintained. Each amends the other. There is no point in creating theoretically perfect systems that are in practice too cumbersome or too slow; eventually they will fall into disuse. Particularly in a fast-changing company, procedures and forms can be overtaken by events very quickly. A well-designed quality document system should be equally fast moving.

As soon as one part of the procedural system falls into disuse, for instance if there are parts of a standard form 'we never use now', or parts of the written procedure that have been superseded by new machinery, the controlled relationship between what is laid down and what is done is lost. It's as if the

CASE STUDY: A PROCEDURE DOCUMENTED

Procedure for writing a procedure	Proc. No. Doc-001
Approvals: ____________________	Issue No: A
____________________	Date: November, 1990
____________________	Page: 1 of: 3

1. *Purpose*:
   The purpose of this procedure is to detail the requirements necessary in documenting a procedure.

2. *General*
   2.1 All procedures must be prepared on standard format paper such as this one.

   2.2 All procedures must have as a minimum the following:-
   - title
   - procedure number
   - issue number
   - date
   - page number
   - purpose
   - procedure itself
   - reason for revision

   2.3 The decision, as to whether or not a written procedure is necessary, should be taken on the basis that if the absence of such a procedure could adversely affect the quality of work being performed, then a procedure is necessary.

   The language used in a procedure should be specific enough to avoid ambiguity and general enough to ensure it is practical.

   The Documentation Control department controls the issue of procedure numbers. A log of these is kept in this department. The numbers are allocated as follows:

DOC	–	XXX	Sequential number
			denotes originating department, e.g. Documentation Control
QAP	–		Quality Assurance
DOC	–		Document Control
MFG	–		Manufacturing
MAT	–		Materials Management
ENG	–		Engineering
FIN	–		Finance
SAL	–		Sales & Marketing
PER	–		Personnel, etc.

3. *Procedure*:

3.1 The Originator will prepare a draft procedure.

3.2 The Originator will then forward the draft procedure to all departments affected by the procedure for their comments.

3.3 Departments are normally given two weeks to respond with their comments (Originator should specify in a covering memo, the deadline for comments and circulation list).

3.4 Comments are taken into account and the procedure modified if necessary. The M.D. will arbitrate on any disagreement.

3.5 The Originator signs and dates the document under the Approvals Section.

3.6 The modified procedure is then issued to the same departments as in 3.2 above for their signature and date.

3.7 The procedure is returned to Documentation Control and kept on file there as the master.

3.8 Documentation Control issues official copies of the procedure (each page stamped in green with wording 'Approved only if stamped in Green') to the relevant personnel as per DOC–004.

Source: IQA

company's steering wheel had too much play in it. Once the rot starts, other parts of the system will begin to be neglected, and daily practice will drift further and further away from the quality system. It is the job of the internal audit function to detect early symptoms of this drift.

However, just as there is no point in establishing elaborate systems that are unworkable, it is not enough simply to codify existing practice. There are formal aspects to a quality system that must be added to daily practice. The trick is to design and introduce these elements with a sensitive eye to times of rush as well as ordinary times. This is best done with the full involvement of key departmental staff.

The following paragraphs describe a method for establishing a documentation system that concentrates on involvement and consensus. The formal side of a successful quality system depends on the documentation; this in turn depends on establishing a set of documents and procedures that the company can live with on a day-to-day basis both in easy times and in hard. It is important that reasonable time be allowed for the creation of these documents; a good job will often take a hard-working team 18 months.

Companies often start this process by writing a company manual. Unfortunately this is like a builder starting with the roof. As we have seen, the manual is not a series of detailed plans and procedures: it is a statement of what the plans and procedures should be. If this distinction is not preserved, the final manual is likely to be an enormous and unwieldy document. In practice it is probably easier to start with plans. As we have seen these are the step-by-step flow charts of the activities necessary to deliver a service or product. Key activities can then be identified as requiring written procedures. Only then should the manual itself be finalised, to embody the policies and understandings formalised while establishing the plans and procedures. The manual should be no more than 30 pages long (Oakland 1989, 161–5).

Stage 1: The first step in drawing up the set of quality management documents is to establish a team with representatives from each department. Each team member will be responsible for consulting with and carrying his or her department's views into the general discussion. This team should be chaired by a senior executive and facilitated by the Quality Manager, or whoever has overall responsibility for producing the document set. An outside consultant may well be a member of the team also.

This team's first task may be to identify and sequence the main activities for each of the products/services of the organization. It is not necessary to do this in great detail, but this overview will form the basis for the establishment of separate quality plans based on flow charts. (A plan, remember, is made up of a sequence of activities and checks that create a product or service. A procedure is an instruction on how to carry out a particular activity or check.)

Stage 2: The next task is to find out what informal quality plans and standards exist for each product or service, and the forms, procedures and checks which are in place for each activity. One of the interesting things about the formal establishment of quality plans for each activity is the discovery that there are always informal, loosely understood plans, standards and procedures in existence. Everyone works to some standard of quality, even if only at the level of 'Ah, sure it'll do.' Quality systems always exist, and are frequently based on different values from department to department. The creative side of this process is to bring such unwritten ideas into the open and into scrutiny.

For each of these activities, the following elements need to be established:

1. What is the sequence of activities?
2. What materials, instructions and other inputs are required?
3. Who does the job?
4. To whom do they report?
5. What control procedures are followed in normal and (especially) abnormal circumstances to ensure that the job is done correctly?
6. What inspection and reporting systems are in place?
7. What standards distinguish a good job from a bad one?

In the process of answering these questions the team will begin to put together a set of quality plans and procedures. This will include standard forms and instructions, detailing materials to be used and sequence of operations. Inspection standards, checks and report systems are part of this package.

Stage 3: The team now has an outline set of quality plans, procedures and reports for each major activity. The next stage is to develop the most suitable of these to a full-bodied form. This means filling in the gaps, drafting, missing specifications or procedures in the department itself, and establishing report formats that monitor key data sets.

The purpose is to set up a model that other departments can follow. At the same time the members of the team are learning how best to go about the whole process of establishing a set of documents that the department and the company can live with. This experience will be valuable when more awkward areas are tackled.

Most of this effort will fall on the Quality Manager and the departmental team member intimately involved; the other team members during this time can continue to build up the information required from Stage 2.

Stage 4: With the model quality plan and procedures established, work can press ahead on other departments. The model will make it clear to heads of departments what is required, but the authority of the senior executive involved may be required to achieve cooperation. Coercion will defeat the ultimate acceptability of the quality system. A more subtle problem arises when departmental representatives, perhaps unused to discussing systems in this way, enthusiastically endorse procedures that turn out to be quite impractical in moments of stress. This can undermine the practical authority of the system. It may be a good idea to have a 'dry-run' of some of the new procedures by envisaging step-by-step how they would operate in typical situations so that everyone can see clearly how things would change.

Stage 5: When plans, procedures and reporting systems are in place for each of the major areas identified at the beginning of the process, work may start on the quality manual itself. Typically nowadays the manual is based on ISO 9000, often to the point of using the ISO 9000 paragraph headings as the basis of the contents. Physically the manual is normally bound separately from the plans and procedures; the way these are presented depends on the company. In some cases procedures may be gathered best by department, in others by product group. This separation of the manual from the procedures goes a long way to preserving the confidentiality of the company's special activities.

Stage 6: The final set of documents, consisting of the quality manual, the plans, the procedures and the report structures, are now ready to be formally approved. This is normally done by a formal signature of the document by the Chief Executive for the set of documents as a whole, and by divisional heads for their own areas. The formally approved document can then be reproduced and distributed as specified. It is desirable that this be done as formally as possible, so as to impress on the recipients the importance of the document.

The minimum set of documents

ISO 9004–2 does not specify a minimum set of documents. However it is clear from the use of certain words and phrases that certain documentary requirements are implicit in the standard.

These include:

1. A quality manual.
2. Quality plans for each service.
3. Procedures for all those operations where the lack of such procedures will affect the quality of the product, particularly where professional or skilled workers are involved. There is often difficulty as to whether a procedure is necessary in a particular case. The best approach to a solution is as follows: the quality documentation specifies the type and level of qualification and special training the worker requires to do this job. If such a person coming new to the job would, by dint of his or her knowledge, perform the task in the prescribed way, then it can be assumed that the specification of a level of training is sufficient. If on the other hand this cannot be assumed (procedures relating to local circumstances, particular customers, etc.) then a procedure is necessary.
4. Records and work instructions. All records, checklists, dockets, etc. that form an essential part of the operating system should be included in the quality documentation.

Document control

Once the set of documents is created, they must be controlled. Fundamentally this means ensuring that they are kept up to date and that everyone is working from the most recent version. Document control is a concept familiar to those

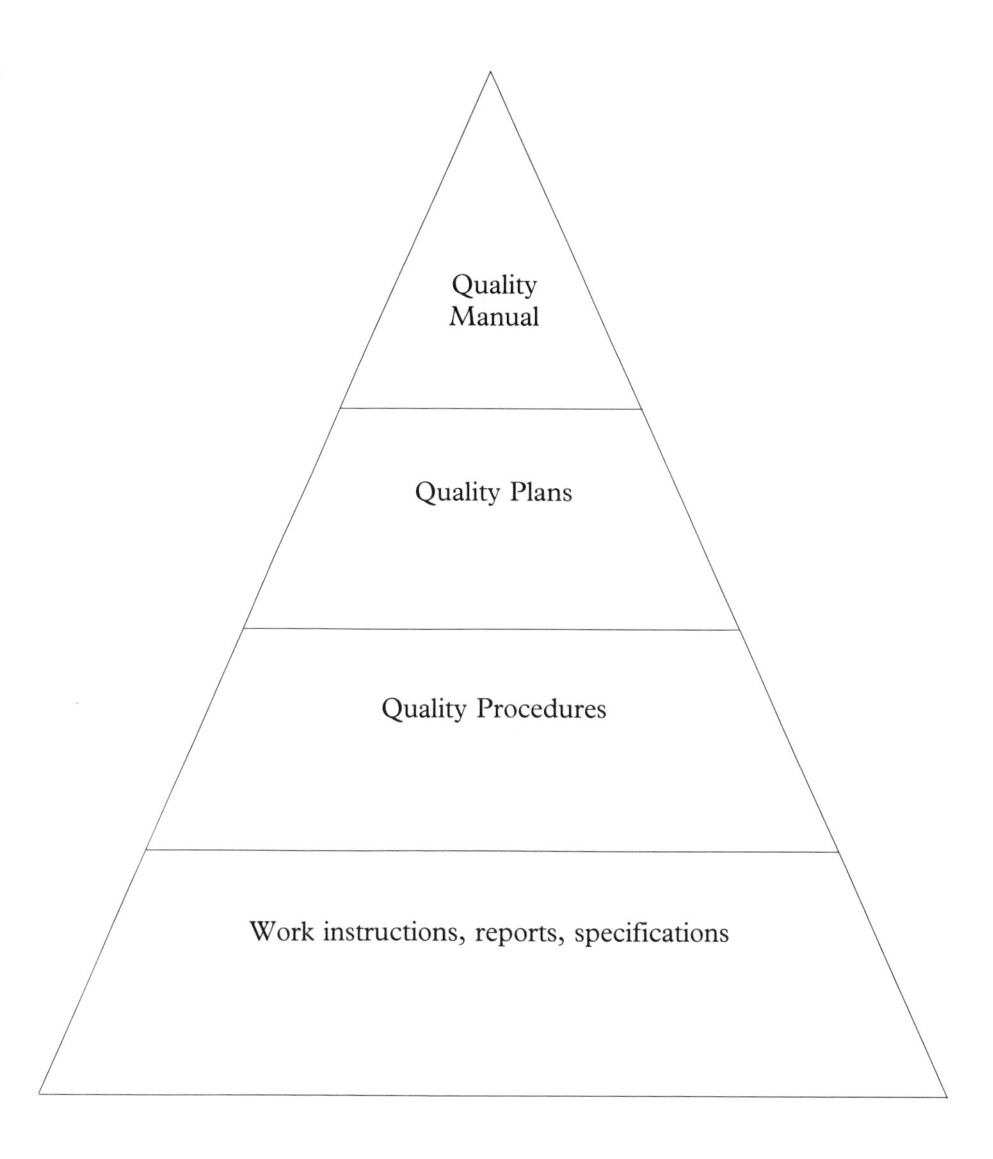

The Quality documentation hierarchy

in engineering and manufacturing, but perhaps less familiar to service industry people.

Its origins lie in the necessity for ensuring that during a large scale manufacturing project sprigget no. 307 will fit neatly into widget no. 887B, even though they are made in different locations. Obviously if the sprigget manufacturer is working from an older set of drawings than the widget maker, the chances are they won't fit. It is therefore important to ensure that the latest version of all drawings is in every critical place, and that no one is working from obsolete or erroneous plans. This is done by a controlled document procedure.

The ISO 9000 standard for quality management systems requires that methods be established to control the issue, distribution and revision of all documents that affect the quality of the service or product. The standard goes on to require that these methods 'should ensure that documents are:

- checked for completeness and adequacy.
- approved by authorized personnel.
- released and made available in the areas where the work is to be performed.
- reviewed for any necessary revision.
- removed when obsolete.' (ISO 9004–2 paragraph 5.4.3.2)

In practice this means that all documents that form part of the system (that is the quality manual, the plans and the procedures) must meet the following requirements:

1. Controlled and documented issue system: The quality manual should lay down how and by whom the documents are issued. Normally it will be required that the key personnel affected by the procedures in the document should be involved in the issuing procedure. They should also be involved in checking the procedure to ensure that it meets the requirements laid down in the manual and that it is adequate to achieve the quality levels specified. The fact that this check has been done should be made clear on the face of the document, normally by an approval signature either on each page or on a prominent early page. A master copy of documents should be kept in a specified place (typically by the Quality Manager).

2. Distribution: Once the document has been drawn up and approved, it has to be distributed. A balance has to be struck between making systems and practice information available to all who need it, and piling every desk with only remotely relevant information. There is no hard and fast rule here.

Some companies make their quality manual available to everyone; others split it into sections and make only those bits that refer to particular areas available in those areas. Perhaps the most popular system is to have the high-level documents such as the quality manual available to the top level of management only, and then establish local procedures and systems manuals for each department. Whatever choice is made will be described in the quality manual.

The revision status of the page and some evidence that the document is complete should be evident on the face of each page. This may be done by a line at the foot of each page as follows: 'Supplier assessment procedure: version 3, June 1990 page 2 of 3.'

This should be backed by a formal list detailing the latest revision date and version of each of the controlled documents, and the distribution list for each. The internal audit should check:

- that this listing is up to date.
- that everyone on the list has a copy of the document.
- that this copy is complete, accessible and in the latest version.

3. Review and reissue: No system or procedure stays the same for ever. The Quality Manager (or whoever has control of the documentation system) will need to establish and document systems for ensuring that technical or specification changes are brought into the system. This might be done by the internal audit. Any new customer requirements, new products or machinery that may change or affect the delivery of quality to the customer will require new procedures to be laid down. Normally this will be done by the same group who wrote the original procedure, and checked in the same fashion. As before, the revision status of the procedure must be clear on the page, thus: 'Supplier assessment procedure: version 4, October 1991 page 2 of 3.'

When the new procedure is written and approved, the fact of the revision is recorded in the list of revisions, and this new list and the new procedure itself must be distributed to the holders. At the same time as the new documents are distributed, the obsolete ones must be recovered — one in, one out. The Quality Manager must establish a system to ensure that all holders have in fact received the new documents and removed the old ones. Typically this will be done in person, perhaps backed by a receipt signature.

ACTION

- **buy a copy of the ISO standard documentation and use it as the basis of a systems audit of your business.**
- **how far does your company meet the requirements of, say, paragraph 4.3.2.2 relating to personnel training?**
- **how far does the company possess standard operating procedures, service delivery specifications and so on?**

SUMMARY

Formal quality systems are the shell inside which detailed quality improvement grows. The international guideline ISO 9004–2 has been written to provide a framework which any service company can use to establish good formal systems. This standard is backed by a regular series of independent audits to ensure adherence to the standards. The purpose of documentation and documentation control is to ensure that all the procedures required to deliver full customer satisfaction are written down in a systematic form. At the ground-floor level this means specifying individual procedures (often called Standard Operating Procedures), and the reports that monitor their performance.

Above that level are the quality plans which control groups of procedures, laying down the methods for ensuring that these groups of procedures achieve the quality targets. Company-wide, the plans are united by the quality manual which describes how the plans are set up, what aims they are to meet and how they are monitored and documented.

Documentation control ensures that the documents that lay down these various procedures, plans and manuals are drawn up in a properly rigorous form, properly authorized and distributed, and kept up to date.

CHAPTER 12

Continuous Improvement

'Even if you're on the right track, you'll get run over if you just sit there.'
Will Rogers

The history of management thought in the twentieth century is littered with brilliant enthusiasms. At the very beginning of the century it was F.W. Taylor's Scientific Management movement that was going to revolutionize the running of businesses; in the following decades, first cost accounting, then management by objectives, operations research and systems thinking, marketing, organizational behaviour management, and many other ideas became fashionable. After the initial flare-up of excitement, most of these ideas became merely part of the general armoury of management disciplines. Perhaps this too will happen to quality thinking.

There is good reason however to think that it might not. Unlike earlier enthusiasms, quality gets to the parts other ideas cannot reach. The quality concept unites production, accounting and service areas into a single focus. Quality starts and ends with customers, and in between provides structures and motivations that unite management and workers in an understandable way. It is this ability firstly to focus the company on its core activities, and secondly to institutionalize dynamic processes that gives the quality concept such power.

These facets of quality activity are tied together by the formal quality system: the network of operating procedures and policies that provide a rigid structure inside which quality grows. The corrective action procedure is perhaps the most important aspect of the formal quality system. It changes the quality system from being a set of administrative procedures and aspirations to a dynamic process.

This difference is critical. What otherwise might be a set of dry-as-dust rules becomes a culture of continual improvement. It institutionalizes change.

Like plants, businesses are either growing or dying. The growing may be slow, even quiescent, but it must be continuous. Translated into business terms this means that the company must continue to improve relative to its environment. The Malcolm Baldrige criteria put the point thus: 'Achieving the highest levels of quality and competitiveness requires a well-defined and well-executed approach to continuous improvement of all operations and of all work unit activities of a company. Improvements may be of several types:

(*1*) enhancing value to the customer through improved product and service attributes;
(*2*) reducing errors and defects;
(*3*) improving responsiveness and cycle time performance; and
(*4*) improving efficiency and effectiveness in use of all resources.

Improvement is driven not only by the objective to provide superior quality. It is driven also by the need to be responsive and efficient — both conferring additional marketplace advantages. To meet all of these requirements, the process of continuous improvement must contain regular cycles of planning, execution and evaluation. It must be accompanied by a basis — preferably a quantitative basis — for assessing progress, and for deriving information for future cycles of improvement.'

Continuous improvement does not of course just happen. It has to be planned for, organized and driven. Like every other successful activity, continuous improvement is a struggle against inertia, against established and comfortable methods, against working solutions. Continuous improvement is in fact a nice way of saying continuous change. It means constantly questioning yesterday's good solutions: moving machines that seemed fine where they were, but just might be better somewhere else; changing the menu, before anyone complains of sameness; not waiting until something is broken before you improve it.

This all goes against our deepest instincts; throughout history change has been seen as something to be dreaded rather than encouraged. The desire for change was seen as an attack on the hard-won status quo. Words such as 'radical', 'enthusiast', 'liberal' were not terms of approval. After all, existing systems had been evolved as solutions to problems faced over time, everyone is familiar with this way of doing things, 'change for change's sake' is easily condemned, and how do we know we are not simply jumping out of the frying pan into the fire? 'Always keep a hold of Nurse', wrote Hilaire Belloc, 'for fear of finding something worse.'

The stimulus to improvement comes from two sources. First of all, there is the corrective action programme built into the formal quality system. This is essentially reactive to particular errors and mistakes caught internally or noted by customers. The simple theory behind this is that anyone can make a mistake once: only a fool makes the same mistake twice.

Secondly, there is a more pro-active force for change generated by special groups set up to find and solve large problems. Typically, they are called Quality

Circles, Quality Improvement Teams, Quality Teams, or Quality Task Forces. These tackle deeper difficulties such as systems problems, problems that span several departments, and problems that are too wide for any single manager to address. There are always more of these in any company than you would imagine possible. In one small and successful printing company a group of managers quickly identified over 20 such deep problems in a brainstorming session.

It is an essential requirement of the ISO 9000 series of quality systems management standards and guidelines that 'whenever a nonconformance is detected, action should be taken to record, analyze and correct it. Frequently there will be two stages of corrective action: firstly, an immediate positive action to satisfy the needs of the customer; secondly, an evaluation of the root cause of the nonconformity to determine any necessary longer term corrective action to prevent recurrence of the problem. Longer term corrective action should be appropriate to the magnitude and effect of the problem. When implemented, the corrective actions should be monitored to ensure that they are effective.'(ISO 9004–2 para. 6.3.5.2)

The guideline suggests a six stage corrective action process. Each stage presents its own difficulties:

1. Detecting the error: The trigger point of the system is the detection of an error. In most cases this will be uncontroversial: if a delivery is late, a letter sent to the wrong client, an order form wrongly filled out, there is no dispute. In service environments where, as yet, product specifications are vague, some activities giving less than full customer satisfaction may be ill-defined. It seems necessary for the quality system therefore to specify in some detail what constitutes a deviation from the specification at each stage of the process.

2. Recording the error: Because much service industry activity is done face-to-face, service deliverer to customer, errors are often spotted and corrected immediately. Persuading staff to record errors can be difficult. On the other hand we know that only a small proportion of customers with a problem bother to complain. Excellent companies go to some lengths to encourage customers to complain.

3. Correcting the immediate results of the error: The first action as soon as an error is detected is to stage what Tom Peters calls 'a remarkable recovery'. For any company that has not yet reached Zero Defects, this concept is essential. Peters quotes a typical case: 'The electricity failed while a woman was at a grocery check-out. When the clerk would not let her take her groceries home, she was irate. She was a regular customer and was rushing to prepare her husband's 50th birthday party dinner. She returned home and, fed-up, she called the store. The manager asked her what she had wanted to purchase. A half-hour later he was at her home with the groceries — gratis of course — along with a birthday cake, inscribed.' (Peters 1986). An ordinary recovery would be to ring the woman back immediately and tell her to collect and pay for her groceries, perhaps making a favour of keeping the shop open after hours. The overkill (free groceries, inscribed cake, personal delivery) is the important point. If you make a mistake, correct it quickly, stylishly and ideally in such a way that your customers will tell their friends.

4. Analyzing the cause of the error: The next step is to see what must be done to prevent the problem occurring again. This operation must start not with solutions but with analysis. In busy companies which put high value on immediate action, managers are pressured into making snap diagnoses in order to rush on to the next problem. Unfortunately this inevitably means that the solutions adopted will not fit the problem, which will flash into flame again in due course. Because no time is taken to examine the problem carefully, proposed solutions are likely to be either superficial or prefabricated, or both. Everyone tends to produce their favourite line of solution — 'more discipline' says the general manager, 'incentive bonuses' says the sales manager, 'cost controls' says the accountant, 'written procedures' says the quality manager. Eventually the company chokes from the smoke of ill-doused problems, still smouldering away. Furthermore, each half-solved problem complicates the solution of the next one.

In order to avoid this managers and operators must take time to allow genuine and specific solutions to emerge from a thorough understanding of the problem. A good way to start this is to analyze every problem in two sessions: at the first, no solutions whatsoever to the problem should be considered, only cause and effect diagnoses. Only at the second should possible solutions be considered.

5. Preventing recurrence of the problem: Interesting and lasting solutions emerge from a good analysis of the problem. Brainstorming and other analytic techniques frequently reveal that the root of the problem is different from what was first thought; just as an acupuncturist seeks an exposed nerve far from the apparent symptom, so business problems frequently originate in unexpected places. The effective solution must be sought at the point of origin. In certain cases it may be necessary to pilot a solution, to check that the true point of origin has been found. Whatever solution is proposed should be written down, and one person with sufficient authority delegated to ensure that it is carried out.

6. Monitoring the effects of the corrective action: Whether a full or only a pilot solution is adopted, it is an essential part of the process to monitor the results. This may require establishing special measuring records. Normally, though not necessarily, the person responsible for implementing the solution will also be responsible for verifying the results.

Improvement

The corrective action process is essentially reactive. Problems, defects and errors are produced, and the process attempts to ensure that they do not recur. The continuous improvement process, on the other hand, is essentially creative. It attempts to drive the business forward. In his bestselling 'management novel' *The Goal*, Eliyahu Goldratt introduced the idea that all operations and systems are prevented from doing as well as hoped by constraints. These may be physical (lack of machinery, space), personal (lack of skill, knowledge) or organizational (we do it that way because we always have). He believes that in practice the most serious and damaging constraints are those based on attitudes, insights or understandings that were valid when they first arose, but which have now lost significance.

The original training novel was expanded later into other books, notably *The Theory of Constraints*. This theory is aimed mainly at production managers in factories, but has relevance to other operations. It says that in any system there are bottlenecks, constraints, and inhibitions that prevent the objectives of the organization being achieved. A fleet of ships must travel at the speed of the slowest to stay together; however fast the frigates and gunboats may be able to go, their speed is limited to that of the lumbering supply ships since they must stay with the fleet. The only way they can go faster (and still remain in the fleet) is for the supply ships to go faster. The speed of the convoy is dependent on the speed of the slowest vessel.

Goldratt generalizes this understanding to all operations, and adds a management implication. Since the convoy can only go faster if the supply ships are speeded up, all management's efforts must be put into that, even at the expense of slowing down the other ships. The whole focus of the management effort must go into maximizing the improvement at the bottleneck or constraint. Real improvement is only possible at the bottleneck, and effort spent anywhere else is simply creating more of a jam at the bottleneck. For instance, one can imagine that the productivity (and so profit) of a sandwich bar is limited by the number of people it can serve during peak hours. All management effort must therefore be put into maximizing peak throughput; no amount of decoration, improvement of the menu, additional products or computerization of the stock control system will deliver any extra profit comparable to attention to that constraint. Indeed, some of these may even be counter-productive: if you increase the menu without increasing the amount people spend on average, thus simply offering more choice, you will simply add to the queue length as people take longer to choose.

Goldratt does not believe that businesses are simply and solely in existence to please customers. This confuses the means with the end. People come together in a business organization to make more money (and have more fun) together than they could singly. It is his view that 'romantic' TQM and JIT writers have been overwhelmed by aesthetic considerations and forget the bottom line. A factory as clean as a pharmacy may be pleasant to look at, but does it increase sales? Just-in-Time techniques, in which the breakdown of a single unit in the flow can paralyze a whole factory may be fine in theory, but in the real world random breakdowns do occur, and some buffer stocks are desirable to prevent negative effects from them. Perfect preventive maintenance is simply not possible. (See Harvard Business Review January/February 1991.)

This is not to say that Goldratt wishes to return to the old days of cost control dominated policies, of accountancy run environments. Quite to the contrary. He is clear that customer satisfaction at the moment of truth is critical to success. On the other hand he rightly suggests that we should concentrate on those improvements to customer satisfaction that maximize also the company's return on investment. In order to concentrate the mind on improvements to the service that will do this, Goldratt proposes three key measures. Any improvement should be directly related to improving one of these three factors. The three factors answer three simple questions 'How much money is generated by our company? How much money is captured by our company? And how much money do we have to spend to operate it?' (Goldratt 1990). These three factors he calls:

- *Throughput*, or the rate at which money is generated by sales; that is, real sales, not possible future ones stored as R & D or in a warehouse.
- *Inventory*, or the money the system invests in purchasing things it intends to sell. This refers particularly to raw materials, but also includes plant, which is turned into saleable products via depreciation.
- *Operating expense*, which is all the money the company spends turning inventory into throughput.

This is not the book to examine the impact of these definitions on the operating systems of any company, nor to go into them in any great detail. They are presented here simply as three key factors that concentrate a manager's mind. If it is not clear that the improvement project will either increase throughput, reduce inventory or reduce operating expense, then why is it being considered? One advantage of these measures is that the good direction is comfortably and intuitively obvious to any manager.

The continuous improvement process then, starts with two thoughts. Firstly that the objective is to change for the better one of the three key indicators; and secondly that this is best done by effecting improvements at the system's constraint points. But how are these to be identified?

First of all it is important to recognize that there are, and always will be, a constraint. Indeed it is likely that there will be more than one, otherwise the business would make infinite profits. Since (we can fairly assume) you do not, there are constraints preventing you. These are the elements whose effectiveness controls the size of your profit; exploit them to the full, and you maximize profit.

Constraints come in all forms: the most profound and effective of which are policy constraints. We have seen from the PIMS data that in the best managed companies, as much as 80 per cent of return on investment is a function of being in the right market. Perhaps it is not practical to suggest that a sandwich bar proprietor should become a computer programmer (a resource constraint is likely to intervene at least), but it is relevant to explore how far pre-set ideas reflect how things are done and why. This concept of understanding the customers' value chain and designing a product as a result is explored more deeply in Chapters 3 and 4. Policy constraints may be of all sorts; they are likely to be strong and difficult to remove. Other types of constraints come from resources, markets, suppliers and internal organizational matters. A cool look at the business will no doubt deliver a list of constraints, their significance for the three factors and your company's success can thus be roughly graded.

The next stage is to exploit the constraint. This means organizing your business so that you make the most of the scarce resource. Put like that it sounds obvious, and it is. Unfortunately, because many companies fail to identify their constraints properly, they also fail to concentrate their management effort into exploiting them properly. This may mean in effect slowing down one part of the operation in order to devote resources to maximize the effectiveness of the constraint.

Only when you have maximized the utility of the scarce resource or constraint operation should you explore ways of removing it. It may even have then ceased to be significant; in which case you therefore return to stage one, and identify another critical restraint area.

How to focus

The key to a continuous improvement process is focus. In an environment where managers are like the man in the circus, whirling 130 plates on the top of poles, dashing from one pole to another to keep all the plates in the air, it is impossible to focus. One plate after all looks much the same as another; at best it is possible to distinguish the plates by their immediate likelihood of toppling, or possibly because the ringmaster has brought one to your attention. The urgent, in this scenario, constantly defeats the important.

Focus is the technique that prevents this from happening. Most managers, if away from their centres of activity, would agree that there are projects that if only time could be found to do them would pay off handsomely in the long run.

The only way this will be done is by establishing action teams that support and motivate each other. These might be called Quality Improvement Teams (QUITs), Quality Action Teams or any other similar name. Their job is first of all to identify a particular problem, using the focusing methods described above, and then to set about evolving solutions to the problem. These teams also have a role in developing the company's problem solving techniques, so that the same systems can be used at local departmental levels. The introduction of the Quality Improvement Team approach is very often done with the assistance of an outside consultant as facilitator; the Irish Quality Association maintains a directory of consultants who are familiar with such work.

However the system is started, the Quality Improvement Team has to become committed to the solution of the problem. This means taking time out from day-to-day activities to devote to information gathering, to brainstorming sessions and to analysis. Unless the group takes the problem, and the possibility of solving it, seriously, the project will collapse at the first difficulty. This means that senior management has to take a strong and continuing interest in the operation, if they do not themselves take part. Ideally of course the company's top people should be the first Quality Improvement Team, especially in the light of the received wisdom that 85 per cent of problems that show up at the moment of truth are ultimately caused by management doing or failing to do something.

Defining the problem

A problem displays itself as a set of symptoms. However, treating the symptoms is not the same as curing the patient. The cause-effect-cause chain must be explored to get to the root of the problem. Suppose a hotel is having problems with making sure that bedrooms are prepared before the new guests book in. The first thought is to check on the staff discipline, then perhaps to their training and motivation, then to their numbers, then to the cleaning staff recruitment, to their expenditure, to the management allocation of budget, and so on. In the end the problem turns out to be one of management and resources, not staff discipline. It might just as easily have turned out to be a problem of information: if the system does not tell the floor staff which rooms to clean, then it is likely that the cleaning cycle will not coincide with the booking cycle.

There are various techniques for helping a group to define a problem. The first is simple brainstorming combined with a cause and effect chart.

How to brainstorm

1. Choose a facilitator, whose role will be to ensure that the process is adhered to.
2. Write the starting point of the problem in the middle of a flip chart.
3. Each member of the group spends a few minutes noting the chain of events which causes the symptom/problem.
4. The group facilitator then asks each member in turn to propose a cause for the symptom; if no thought occurs, then the group member simply says 'Pass', and the leader moves on; absolutely no idea is too simple or too basic to be introduced at this stage, and none of the contributions may be commented on during this process.

CASE STUDY: ULSTER BANK QUALITY SERVICE ACTION TEAMS QUALITY INITIATIVES

Reorganisation of Cheque Book/Card System
Develop more big business in branch
Improved Car Park Facilities
Improved Counter Service
Creation of Product Knowledge files, talks, indexes, competitions
Improved Stationery System
Improved Foreign Till Facilities
Reduced Queues
Improved Quality Service over lunchtime
Initiated Opening Account Pack
Improved Appearance of Office
Improved Awareness of Staff/Customer
Improved Aide Mémoire update
Office Re-organisation
Customer surveys Introduced in Branches
More information on Customer Records
Improved turnaround of School Savings Scheme
Improved training among staff
Kept up superior service during renovations
Organised Move to new premises
Re-organised Standing Orders and Direct Debits
Improved Appointment System
Organised Open Days
Customer Suggestion Box Created
More Local Advertising
Improved atmosphere in branch
Improved response to telephone calls
More Henry Hippo promotion
Update of Rates Board displayed in banking hall
Installation of Servicetill at Branch
More Statement Detail
Improved Facilities for disabled customers
Increased ATM Network
Use of Name Badges
More small business visits
Bandit screens removed
Increased Privacy
Introduced Office Flowers/Hanging Baskets/Window Boxes/Plants
Use of In house Display/Local Artists
Introduced Fast Lodgment Box/Bulk Lodgments
Introduced Call and Care System
Created Guide to Health & Safety
Introduced Staff Business Cards for Customers

Source: Ulster Bank

5. When the first set of ideas has been produced, some of the participants will see connections or affinities between causes, or causes of causes; if this is the specific objective of the session, then the ideas (perhaps written onto post-its) should be grouped on the board.
6. The leader goes round again, and again, and again, letting ideas build on each other until everyone is finished.
7. It may be fruitful now to spend some time individually considering the chart already built up, perhaps between sessions, and to jot down further ideas.
8. If this is done, the facilitator goes through the process again.

At the end of this process there should be a rich flowering of ideas regarding what has caused the particular symptom. In complex situations it may be helpful to copy down the chart and to circulate a fair copy to the group members so that a further session can elaborate on the first attempt. At the end of this process, the brainstorming sessions will have teased out connections or implications that were not immediately apparent or generally understood. By this time the wide ramifications of the problem are apparent, and the group is now in a position to define its problem (rather than simply as before noting the symptoms), and to identify objectives.

A serious problem

Company M, selling a JIT service of 60 or 70 deliveries a week to five major customers, discovered that it was spending £250,000 a year out of a turnover of £5 million on overtime. A group of managers, including the accountant, the quality manager, the works manager and the sales director decide to explore this phenomenon, with a view to cutting down, or at least justifying the expenditure.

1. They start with the proposition that the company is spending too much on overtime. This is clearly a symptom. The brainstorming session produces the basic idea that there are two categories of overtime: that caused by institutional factors (union agreements, slack controls etc.) and that caused by demand for the company's product *now* being greater than capacity *now*.
2. Which is the more significant group of factors? Information from the payment records reveals that institutional overtime is less than 10 per cent of the whole. So the group must explore the second group of factors.
3. Another brainstorming session produces more areas about which information should be sought. Obviously the firm does not want to reduce overall demand, but could they even it out over time, thus allowing the work to be done in normal time, or is it that there is a genuine problem of understaffing? On the capacity side, could it be that workers are stringing jobs out to get overtime (does overtime mysteriously rise just before holidays for instance?); are work practices or job and workforce

planning systems inefficient? Are quality standards unnecessarily high, thus demanding more time from the workforce to meet them?

Clearly in this case the group needs to start gathering facts immediately.

Gathering facts

Many companies have extensive fact-gathering processes. Whole departments are devoted to producing weekly, monthly, quarterly, annual reports and analyses. Unfortunately it will nearly always turn out that when a QUIT group asks a specific question (such as what proportion of overtime is caused by institutional as opposed to demand factors) the answer is difficult to establish. This is because of the difference between data and information.

Data is raw lumps of fact, information is the answer to a question. In the accounts system, the list of entries in the cash books and ledgers are data: the profit and loss account is information, designed to answer one quite specific question about how much profit was made. Data is unusable until it becomes information by answering a specific question. Thus if I ask the travel agent how best to fly to Atlanta, it is no answer for him to point me in the direction of a set of airline schedules. On the other hand if he tells me fully and precisely how to get to Atlanta, that answer is useless if I decide after all to go to Rome. (Goldratt 1990, 79–85)

All existing data-gathering systems are actually already organized to answer specific questions. These may be unstated and even traditional questions, such as, what profit did we make last year? Unfortunately as we have seen, the more precise the answer, the less useful it is likely to be as a response to a different question. Companies try to compromise by providing certain types of general information, but this unfortunately is as useful as the set of airline directories.

So it is likely that our QUIT group will have to devise specific information-gathering systems of its own, and to put these into action. Equally likely, the QUIT will find that one set of information simply gives rise to more questions as the process of fact finding and analysis goes on.

Analyzing the facts

The loop of asking questions, gathering data, looking at the data, asking more questions and gathering more data can be a long one. And it should not be cut short by impatience or by a group member who has a pet solution. Managers who prefer activity to thinking are inclined to rush to one solution and support that through thick and thin. The critical question is: we may believe something to be the case, but can we *prove* it? The act of proving is essential to drive the analysis and to fasten the solution on firm ground. The very fact that we have a problem that would presumably have been solved in the ordinary course of executive action if our standard assumptions were right, proves that they need to be examined.

There are various basic analytical techniques that are particularly helpful in squeezing information out of data. The Japanese call these the seven problem solving tools. They are an essential part of a modern manager's tool-kit. These seven are: the flow chart, the check sheet, the Pareto (80/20) diagram, the cause and effect diagram, the histogram, the control chart, and the scatter diagram. Used in conjunction, these seven techniques form a powerful battery.

Flow Chart: The first task in facing a problem is to find out step by step what actually happens in practice. This is not always as easy as it might seem, and is very likely to be subtly different from the Standard Operating Procedures (SOP). In particular the various exceptions and unofficial sub-routines may be difficult to pin down. For instance, on normal days, your parcel is dispatched with all due process; if there's a rush on, however, the paperwork may in practice be left till after the deadline. Result: errors in documentation. Or the SOP may require that all boxes be weighed to confirm quantities; workers responsible for weighing may assume that this does not apply to very small quantities, or to boxes delivered by a supplier. Result: miscounts in deliveries.

Check sheet: Once you know exactly what happens, you can start to analyze it. A check sheet is a simple form (see illustration) that notes case by case where/why/when a problem occurs. Simple cases include the defect location check sheet. In this the various error locations are listed as they occur and the number of each occurrence is noted.

The most famous use of a check sheet occurred in the great London cholera plague of 1849, when John Snow marked on a map the number of cases of death by cholera in a particular slum area. He noted that deaths tended to cluster round a certain pump; it was closed, and the death rate slumped.

Pareto chart: Having identified the process, and the location of defects, the next step is to put them into order of seriousness. It is well known that a large proportion of defects come from a small proportion of causes. For instance 80 per cent or so of charity income comes from 20 per cent of sources; most motor accidents are similarly caused by a small proportion of drivers; in most companies, the bulk of revenue comes from a small proportion of customers or products. The problem is to distinguish the 'vital few' from the 'trivial many'. Two techniques can be used when faced with a check sheet such as the one above. The first would be to put some kind of value (typically, cost) on each defect. This might give you a table as follows:

The Seven Problem Solving Tools

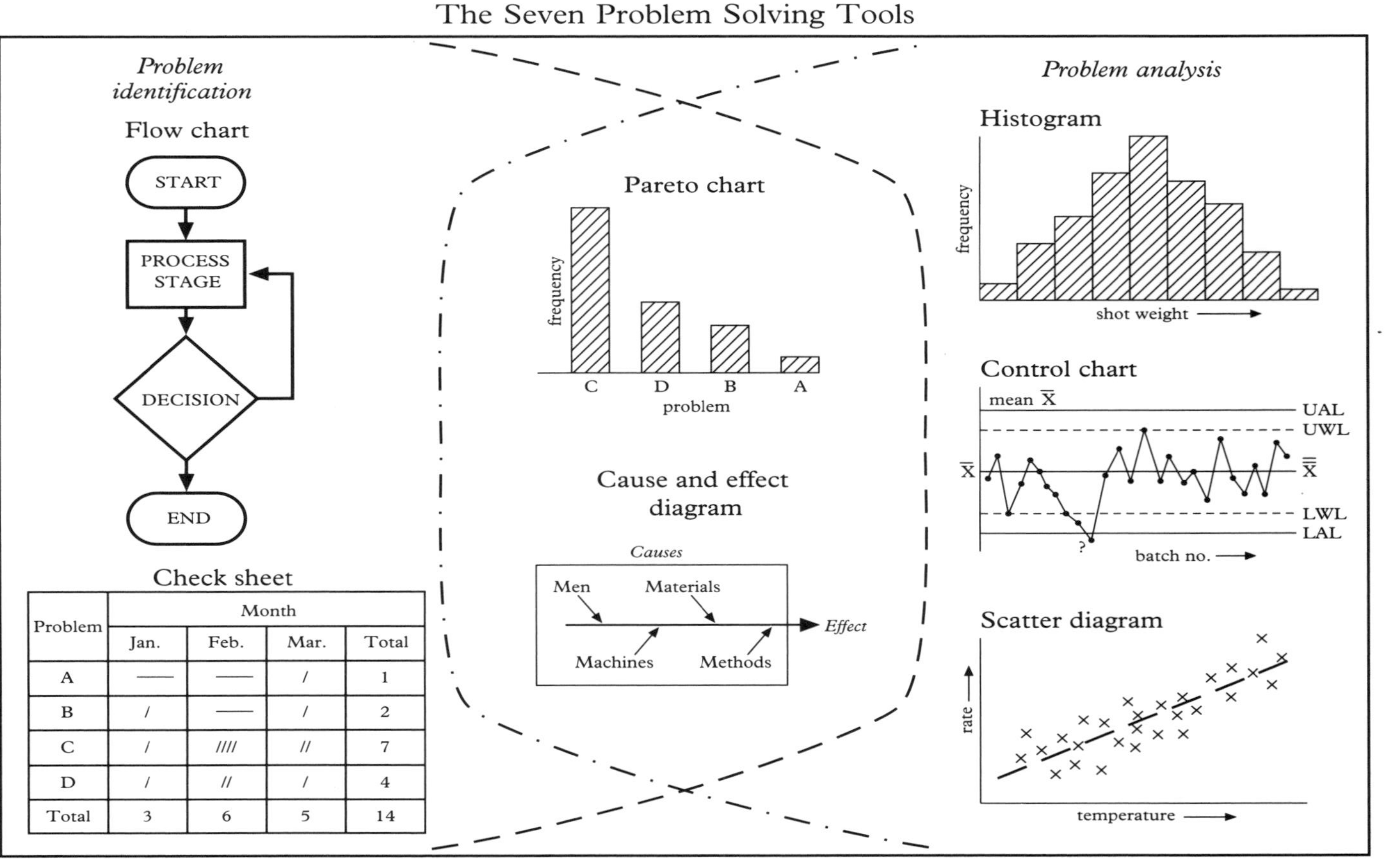

Problem	Month			
	Jan.	Feb.	Mar.	Total
A	——	——	/	1
B	/	——	/	2
C	/	////	//	7
D	/	//	/	4
Total	3	6	5	14

Source: *Quality Assurance* June 1989

Pareto analysis

Defect: Faulty invoice

Location	Occurrences	%	Cost (p)	Value	%
wrong items	4	8	30	120	9
wrong quantity	8	16	20	160	12
wrong price	15	20	15	300	23
wrong extension	3	6	20	60	4
wrong VAT	1	2	45	45	3
wrong name	9	18	10	90	7
wrong address	7	14	75	525	40
other	3	3	5	15	1
TOTAL	50			1315	

In our example we discover that the cost to the company of wrong address (invoice returned to sender, effort spent to trace correct address, etc.) means that the 14 per cent of times that this problem occurs translates into 40 per cent of the cost. Clearly this is the first target. In fact this Pareto analysis shows that making sure that the price and the address are right will remove 63 per cent of cost of defects.

It may not be practical or sensible to assign costs against each defect in this way. In that case group the defects into categories. For instance the price/quantity category accounts for 36 per cent of occurrences, and the name/address group 32 per cent. These are obvious first targets. Note how adding a value element sharpens the analysis considerably.

Cause and effect diagram: The next step is to explore how it is that the wrong address gets on to the invoice. Since this is happening quite regularly, we can assume it is a non-trivial problem, probably with a number of contributory causes. The various causes should be explored in a brainstorming session, and a cause and effect chart drawn up. Otherwise known as the fishbone chart, this displays the individual causes of a single effect into a single complex picture. This and other devices enabling ideas to be connected in this way are an extremely important and valuable part of the process.

If multiple causes are enumerated, it may be necessary to go back to flow charts, check sheets and Pareto analysis to define and evaluate more clearly what is causing the problem. Three statistical techniques may also throw light on it.

Graphs and bar charts: It is said that a picture is worth a thousand words. This is certainly true in data analysis. Bar charts (histograms) in particular enable one to see the 'shape' or dispersion of the data. The eye is better at interpreting pictures than tables, so charts can display hidden regularities and rhythms in the data that can provide clues to causation. Modern spreadsheet programs such as Excel and Lotus 1-2-3 enable users to create bar charts and other graphics very easily. Pareto analysis is often presented in bar chart form.

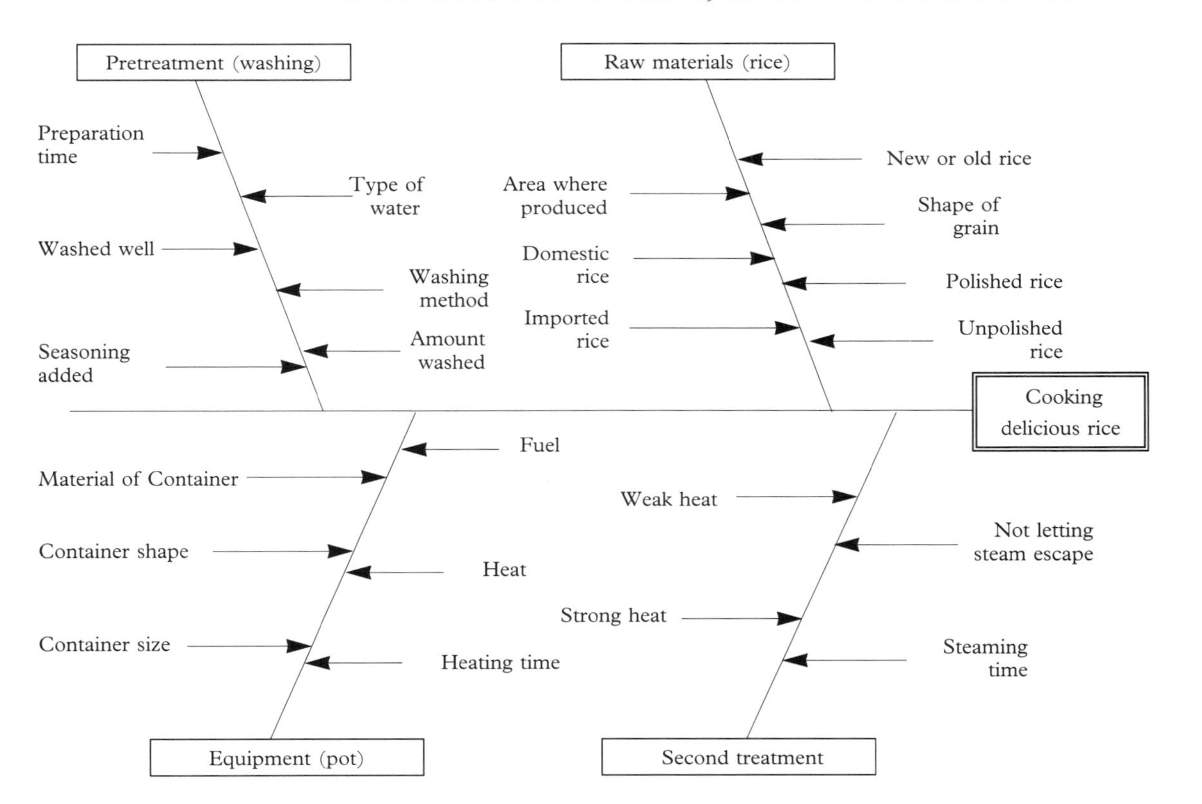

Source: K. Ishikawa *Guide to Quality Control* 2nd revised edition 1986 Asia Productivity Organisation, Unipub White Plains New York

Scatter diagrams: These are graphs that explore the effect of one outcome on another. A publican might notice that the hotter the weather gets, the more lager is sold. A scatter diagram relating temperature to sales will enable him to explore the relationship more closely. In this case he would expect sales of lager to rise with every degree rise in temperature: this is called positive correlation between temperature and lager sales. Negative correlation is where sales of, say, hot whiskey go down with temperature; another example might be absenteeism reducing with age of worker. In these cases there is a direct (and quantifiable) relationship between the independent variable (temperature, age) and the dependent variable (sales, absenteeism). After analyzing the figures (using the mathematical techniques described in the chapters on regression and correlation in basic business statistics textbooks), the publican may be in a position to say for instance that 80 per cent of the increase in lager sales is related to increases in temperature, and that for every five degrees increase in temperature, a certain extra number of barrels will be sold in a week. This is obviously extremely useful for regulating stocks to ensure that hot and thirsty drinkers will get the service they desire.

The data however may reveal that there is no significant relationship between the two variables. Coincidences being what they are, it may also, awkwardly, apparently reveal a relationship where none exists. This happens unexpectedly often. The most famous example occurred when a Scandinavian statistician demonstrated an almost perfect correlation between the annual birth rate in a certain village and the number of storks nesting nearby! In a more prosaic example, a high correlation was found in company M referred to above between expenditure on overtime and the inflow of new orders eight weeks afterwards.

Control charts: These are widely used in industry to check that a process is in control. They assume that minor variations are endemic in any process; these are just 'noise'. Major variations however (outside the range of the minor fluctuations) are undoubtedly caused by a change in conditions. The problem is how to distinguish major and minor blips. Control charts do not address the question as to whether the natural process variations are acceptable. Their sole task is to monitor the operation over time to ensure that operating conditions remain the same.

Control charts do this by setting an upper and lower limit (often related to the historical dispersion of the data) on the chart. As the process goes on, samples of the product are tested and the average or the range is plotted on the chart. In this way trends can be spotted if the process runs out of control.

Control charts can be used in office and service environments to demonstrate the stability or instability of the process. For instance it would be possible to plot the defective proportion of sandwiches/invoices/insurance policy quotations etc. over time.

Action, at last

All this analysis will have given the QUIT a very clear analysis of the problem, based on step-by-step proof. The team will incidentally have gained simultaneously a series of new insights into the real operation of the business. The question of what to do about the problem is most likely to evolve very easily from the solution. If it has not, two possibilities occur. Either the team has been limited in the resources available to produce a realistic solution, or insufficient work has been done in the earlier stages. In the former case, at least the team has a proven basis on which to request further resources, and in the latter, the incentive to do more work.

Once a solution has been evolved, it has to be 'sold' to the staff concerned. This is probably best done by a mini-presentation, showing how the facts were analyzed, and the solution derived. This will have the educational effect of drawing the staff into the analysis, spreading the insights achieved, and reassuring them that the solution is firmly based. It may, embarrassingly, produce considerations that render the solution inoperable. All that can be said about that is: better that it be killed in discussion than damage the company by failing in practice. Hopefully however the group will have considered sufficient input from the staff to ensure that this does not happen.

It may be practical to introduce the solution proposed on a test basis. Obviously this is desirable if it can be done. The test should be monitored over a number of weeks (to allow the novelty effect to wear off) and then if successful, generalized. It is important to establish monitors (such as control charts) that continue to check that the solution is a good one for some time after the general introduction. It is likely that these will be evolved during the information gathering and analysis stages.

ACTION

- **identify those errors that occur again and again.**
- **how can they be prevented from recurring?**
- **identify operational bottlenecks in your system.**
- **use brainstorming and other techniques to minimise its impact.**
- **establish a Quality Action Team to choose and attack key problems.**

SUMMARY

A philosophy of continuous improvement is the basis of the quality revolution of the 1990s. Improvement comes from two sources: the corrective action programme, which is part of the ISO 9000 standard, and continuous improvement, which is typically based on action teams tackling endemic problems. In the first case, the action has two dimensions, the immediate hot-cure, and long-term analysis to prevent recurrence. In the continuous improvement case, the company attempts to focus on key strategic issues such as throughput, inventory and operating costs, and to address these. The seven problem-solving tools (flow charts, check sheets, Pareto analysis, cause and effect charts, bar charts, scatter diagrams and control charts) are used in a defined order to attack problems.

Appendix: implementing TQM at Telecom Éireann

Successful implementation of the Total Quality programme presents a particularly demanding challenge in a large and geographically dispersed enterprise. The task is even more complex when the market-place is rapidly changing and when the long-term intentions of regulatory authorities, both domestic and international, are unclear. That is the challenge Telecom Éireann (T.E.) has decided to face. After a thorough preparatory period during which much was learned about the theory and more important, the successful practice of TQM, T.E. has now embodied a company-wide implementation.

For Telecom, the most fundamental point is that TQM is not a project or a process. It is a journey in a new way of managing that constantly challenges and shapes/reshapes its view of the market and its view of the future. So, T.E. defines TQM as managing the requirements (including those of internal customers) and where every individual function, activity, discipline and level is engaged in understanding and meeting those requirements. It is a proven, disciplined, customer-driven process of change. It involves employees at every level and focuses them on improving all aspects of work by using a structured approach to teamwork and problem-solving.

Focusing on quality opens up opportunities for:

- achieving superior levels of service and customer satisfaction (which are linked to improved financial performance).
- reducing the cost base progressively by establishing a 'Right First Time' ethos and a 'Continuous Improvement' approach to all aspects of work.

Detailed implementation in T.E. will be based on six principles which are at the heart of the approach advocated by PA who are the Company's quality consultants. These are:

- Customer First: The external dimension
- Constancy of Purpose: The leadership dimension
- Focused Involvement: The people dimension
- Act on Facts: The knowledge dimension
- Process Emphasis: The systems dimension
- Continuous Improvement: The learning dimension

The structured approach to problem-solving which will be used by the Company is supported by three pillars: the problem-solving discipline, teamwork and facilitation.

The problem-solving discipline (PSD) has six steps:

1. Define the problem Data
2. Analyze for root caused Collection
3. Generate solutions

4. Plan and implement chosen solution
5. Measure achievement
6. Standardize (if chosen solution has worked)

The six steps are not, of course, watertight compartments. In particular, the first three may need to be used on an iterative basis as data collection proceeds if a team is to be sure that step 3 yields the optimum solution.

The principal tools and techniques used are:

- Brainstorming
- Cause and effect diagrams
- Check-sheets
- Graphs, histograms, bar charts
- Scatter diagrams
- Pareto analysis
- Consensus reaching
- Flow charts
- Affinity diagrams

Using the PSD enables a team (at whatever level) to systematically remove failure from any given process.

One of the principal barriers to using the PSD is people's faith (particularly managers) in their own expertise and experience. 'Why bother with all this data collection analysis and the rest of it? I know what to do' is a common and understandable objection. The whole point of the PSD is to make sure that a problem is not solved on a once-off basis while leaving the root cause intact. Another significant barrier is the fact that, for the most part, the tools and techniques are simple — sometimes to be even childish. But, despite their effectiveness, they are not the way we usually solve problems. There is thus a psychological barrier to be overcome.

Everyone (including managers) must learn the PSD and use it consistently and rigorously throughout the organization. It cannot work piecemeal; it cannot work unless it is fully and universally understood; and that will not happen without everyone receiving the appropriate training and applying it in teams.

Effective teamwork is the second pillar essential to TQM. So the Telecom approach puts considerable emphasis in training on explaining and practising key interpersonal skills — listening (for many people the most difficult), questioning, clarifying, constructive argument, building, summarizing, involvement and recognition — and on teamwork dynamics. It is important that everyone understands and feels comfortable with the phases through which a team may typically pass — forming, norming, storming, performing and dorming.

The third pillar — facilitation — is an important support to teams. Facilitators, who must themselves have good interpersonal skills, play a vital role in a team's early efforts by helping with training, by ensuring they adhere to the problem-solving discipline and by using these skills to help the team through the early uncertain weeks or, in some cases, months.

The management challenge

TQM, of course, is for many people a totally new way of working. It can therefore be unsettling, even painful. That is why top management understanding and commitment is so essential. Five things are needed:

- a willingness to acquire a thorough understanding of the principles and practice of TQM.
- a willingness by managers to help in imparting that understanding to their staff.
- a willingness to bear the initial pain as well as the more enduring exhilaration of implementing the TQM approach.
- a willingness to recognise and support individuals and teams in learning and practising TQM.
- a willingness to give time to make these things happen.

They can be summed up in the phrase: learning, leadership and time.

The T.E. wave process

Experience with large organizations shows that it is usually too demanding to implement TQM company-wide, with the whole organization starting on the same day. Simultaneous implementation on such a broad front would overload the team of internal and external quality practitioners who play an important role in supporting the early stages of implementation, particularly with regard to training and guidance on how to plan and manage the process as it develops. The preferred approach is to implement in a series of moves, each consisting of a number of units.

For each unit within each wave, implementation takes place in four phases. (The durations shown are broad indicators.)

- development and preparation (about 6 weeks).
- commitment and planning (about 6 weeks).
- implementation (about 10 months).
- audit and preparation for the next wave (about 1 month).

In Telecom, the first wave has about a dozen units and covers about 20 per cent of staff.

Other elements planned in the T.E. approach are effective internal communications, appropriate recognition systems and ongoing briefing of the trade unions who have been fully supportive of the initiative to date.

For T.E., as for any other Company, quality improvement is a long haul. It is not easy and there are no miracle short cuts. The idea of continuously improving everything, a bit at a time, is less exciting than making the great leap forward. Yet while the leaps can offer an invaluable contribution to a successful TQM initiative, its steady methodical day-by-day process of using teamwork to eliminate failure and error and achieve continuous improvement is the bedrock of long-term, sustainable success.

CHAPTER 13

Measurement of Quality Costs and Quality Activities

'Quantities derive from measurement, figures from quantities, comparisons from figures, and victory from comparisons.'
Sun Tzu *The Art of War*

Counting, measuring, recording and comparing results against targets are *the* key techniques for the continuous development of quality. The use of measurement in this way is one of the most important of the strategic focus points for quality management, and one of the least used.

The guidelines for the prestigious Malcolm Baldrige Award, managed by the US National Institute of Standards and Technology, an agency of the Department of Commerce, identifies 'Actions Based on Facts, Data, Analysis' as one of its six key indicators of quality management. They go on to describe in some detail what they mean:

> Meeting quality improvement goals of the company requires that actions in setting, controlling, and changing systems and processes be based upon reliable information, data, analyses. Facts and data needed for quality assessment and quality improvement are of many types, including: customer, product and service performance, operations, market, competitive comparisons, supplier and employee-related. Analysis refers to the process of extraction of larger meaning from data to support evaluation and decision making at various levels within the company.
>
> Such analysis may entail using data individually or in combination to reveal information — such as trends, projections, and cause and effect —

CASE STUDY: THE SIX SIGMA QUALITY CONCEPT

Collections of similar items, people or objects show variation among items in the collection. We can observe this natural or normal variation in all of our everyday lives, e.g. height of people/trees, rainfall distribution in countries.

Most such collections fit a pattern called a 'normal distribution' or a 'normal probability distribution'. This distribution has a very useful characteristic which allows us to predict the value or percentage of population that will fall between certain points. It can be fully described using two key measures. The first is the average or mean (called x bar) which is usually designed to be as close as possible to the nominal or desired outcome, i.e. what the customer requires.

However as this is an average, there is variation about this mean value. This variation can be measured by what is called a standard deviation or 'SIGMA'.

So Sigma is a measure of variation . . . and since variation often means deviation from customer requirements, SIGMA can be used as a measure of non-conformance from some agreed requirements.

One characteristic of this normal distribution is that by mathematical derivation it can be shown that 68% of the distribution will be +/–1 Sigma from the mean value, 95% will be within +/–2 Sigmas from the mean.

This relationship can of course now be used to express the percentage that will lie outside these limits, i.e. a +/–1 Sigma process would have 32% non-conformances, a +/–2 Sigma process would have 5% non-conformances or defects. If we continue this we will find that a process or activity at +/–6 Sigma would have .0000002% defects or 99.9999998% defect-free work.

Now unfortunately very few things in this world remain static, there is always a level of change even in the best-controlled processes. So to take account of this fact the definition of SIX SIGMA QUALITY CAPABILITY has been written to allow for some level of shift in the average or mean of a process. Typically this shift can be up to 1.5 Sigma of the process.

This means that SIX SIGMA QUALITY is defined as 99.99966% defect-free work or 3.4 defects per million, which, as was discussed previously, is a very low defect level.

A 3 Sigma process falls 99.7% within the specifications. We used to think that this was fantastic. Well, judge for yourself. If your heart operated at this quality level, it would stop for more than one hour per month. Still fantastic? Many things in the world run between 3 and 4 Sigma. If we translate a 99% defect-free quality into everyday life in the US, this would mean that:

- 20,000 pieces of mail get lost every hour.
- 5,000 incorrect surgical procedures get done every week.
- 200,000 faulty pharmaceutical prescriptions are issued every year.
- and in IBM, it would mean 10 defects in every 1,000 solder joints on a computer motherboard.

Source: IBM

that might not be evident without analysis. Facts, data and analysis support a variety of company purposes such as planning, reviewing company performance, improving operations, and comparing company quality performance with competitors.

A major consideration relating to data and analysis in connection with quality system development, competitive performance and continuous improvement involves the creation and use of performance indicators. Performance indicators are measurable characteristics of products, services, processes, and operations the company uses to evaluate and to track progress. The indicators should be selected to best represent the attributes that link to customer requirements, customer satisfaction and competitive performance as well as to operational effectiveness and efficiency.

A system of indicators thus represents a clear and objective basis for aligning all the activities of the company towards well-defined goals and for tracking progress towards the goals. Through the analysis of data obtained in the tracking processes, the indicators themselves may be evaluated and changed. For example, indicators selected to measure product and service quality may be judged by how well they correlate with customer satisfaction.

The basic purpose of figures in management is to enable meaningful comparisons to be made. If we want to find out whether such a result is better or worse than another, counting and measuring in effect puts the two results into the same scale, for easy comparison. If we can't discover such a common scale, then comparison becomes meaningless, like asking is a poem better or worse than a cloud. Equally, figures by themselves, outside a comparative context, are meaningless. In the *Hitch-Hiker's Guide to the Galaxy*, the mighty computer Deep Thought eventually produced the answer to the ultimate question of Life, the Universe and Everything. The answer was 42. This was not helpful.

At the very simplest level, this is the trick performed by percentage. Is 17/27 greater or smaller than 6/10? The answer only becomes obvious when both fractions are converted to percentages, and we see that 17/27 equals 63 per cent, and 6/10 equals 60 per cent, so the former is the greater. On a more complex level, we try to convert complex results of work into a single scale, so that they may be compared. There are gains and losses in this technique: richness of information is suppressed in order to achieve comparability. We have seen in an earlier chapter that the balance sheet and profit and loss account do not by any means represent all there is to a company, and similarly speed of turn-round in a bank queue may not represent all there is to a customer's happiness.

Counting helps to put things into perspective. A customer going to a hairdresser may spend £12 a month or so; £144 a year. If the worst happens, and she stamps out in a rage, vowing never to return, that does not sound like much to lose, and after all we all have bad days. On the other hand, if good customer loyalty is built up by good service, this will last for ten or fifteen years, which increases the revenue from that customer to £2,000. She will tell friends, say five or ten of them, each of whom represents a similar stream of income. Without too much arithmetic, the loss of that customer can cost £10,000 in lost sales.

In another example, in 1989 the Ulster Bank discovered through a customer survey that although 69 per cent of customers who took the trouble to fill in the survey form had no complaint at all, 9 per cent would be hesitant about recommending the Bank to their friends. As the survey report put it: 'in the Northern Region, this equates to 15,680 and in the Southern Region 24,200 people all actively sending their friends and colleagues elsewhere. This cannot be good.' (See Ulster Bank *Can you spare a moment?*: customer survey 1989.)

In quality management, as in physics, the ultimate object of measurement is to understand and then control real world values; in our case, the service given to customers. The ability to measure and quantify significant results enables real control to occur. Just as physicists have done over the generations, quality managers must first analyze the key elements of what they want to control, and then devise suitable measurement systems. The traditional way to achieve high standards is through endless practice: after a seven-year apprenticeship, and another ten or twenty years journeyman effort, a gifted craftsman, working with a revered master, might achieve extraordinary results. That is how the great cathedrals, the great silver-work, the great illuminated books of the past were created. The pace of change today is such that we do not have the time to master these crafts, nor the space in which to make mistakes. (No draft copies of the Book of Kells survive, but it is known that medieval cathedral towers from time to time fell down soon after erection.)

The modern way to shorten the practice time is by analyzing the information into countable quantities, then counting, measuring and recording results, and building on those results to achieve improvements. This is the scientific approach that has made such an enormous difference to the world since it came out of the laboratory during the Industrial Revolution of the late eighteenth century. The patient weighing of chemicals by wealthy eccentrics in earlier centuries (themselves resting on the doomed speculations of the alchemists) had eventually led, via Dalton's great synthesis of the atomic theory, to great industries producing dies, fertilizers, acids and a thousand other chemical products.

The process of course goes on, increasingly inside industry itself. All sorts of processes may be improved by this approach. An unlikely example is McDonald's french fries, which are probably more popular than any other McDonald's product. It is estimated that the company spent years of research and over $3 million in the 1950s perfecting its method of cooking french fries, a dish that had been popular for hundreds of years. As well as regulating the type of potato, the percentage of solids (21 per cent), and the length of curing the potato, they explored in laboratory detail what actually happens to a potato when it is immersed in fat.

Eventually they came up with a formula: when a batch of cold wet potatoes is immersed in fat at 325 degrees F, the temperature of the fat will drop dramatically. After a while it will begin to recover. When it has reached three degrees hotter than the low point, the potatoes will be perfect (Love 1988). This ability to turn an apparently well-understood product, the french fry, into a precision product has become perhaps the single most important factor in McDonald's success. It is a classic case-study of the importance of measurement and the scientific method in business.

What to measure: costs

Managers traditionally look first at costs. Unfortunately because cost reporting by activity rather than function for service businesses is in its infancy, many service managers are unable to make the best use of financial data. Most accountancy systems measure the expenses in service organizations by functional categories, but make no attempt to attribute them to the activities that cause demand for resources such as marketing, research, purchasing etc. Accountants are currently working on systems to explore the cost of service more deeply. Any group of workers is in practice meeting a demand; activity presumably varies with that demand. It should therefore be possible to measure the demand and the response to it.

Every worker performs certain functions in the company, and has a direct and overhead cost of a certain amount. This enables each function to be separately costed, and ultimately for a 'profitability by service customer measure' to be defined, depending on the mix of products being sold to the customer. We have seen that each service is the climax of a train of transactions, the last of which is the 'moment of truth'. At this point all the notional added value along the way becomes converted into reality (Cooper and Kaplan 1991, 466–71).

A second approach to measuring costs is the quality cost route. The bottom line in all corporate activity is profit. During the 1960s and 1970s more and more companies became aware that market share is the key to profitability. The 1980s have shown that one factor above all others — quality — drives market share (Buzzell and Gale 1987, 103). This is now commonplace. What is less commonplace is the financial analysis of the factors relating to quality — the so-called quality cost. The concept does not appear in conventional accounting texts. This is not because it is an insignificant quantity. For manufacturing companies, it is estimated that quality costs may amount to as much as 20 per cent of sales; in service companies they probably amount to more than double that. In the early 1980s, the Continental Illinois Bank estimated its cost of quality at 37.3 per cent of operating costs, a level by no means exceptional in service industry.

This quality cost is made up of four separate factors. These are:

- prevention costs, the costs of trying to prevent mistakes being made.
- appraisal costs, the costs of checking whether mistakes have in fact been made.

- internal failure costs, the costs of putting right mistakes discovered in-house.
- external failure costs, the costs of putting right mistakes discovered by the customer.

This four-part analysis of quality costs has been used for many years in manufacturing industry, and is strongly endorsed by various quality gurus, especially Dr J.M. Juran. Unfortunately, the categories are not always as self-evident as the textbooks imply. The problem is that prevention and appraisal costs in particular often cannot be distinguished from production costs even in theory. Operations such as on-job training, design, machine make-ready or supplier appraisal are part operational, part quality (Masing 1990). In the end each company has to make some more or less arbitrary decisions about what to include in these categories. Once this is done, period-to-period comparability is assured, though cross-company comparability may not be. It is notable that relatively few Japanese companies make use of the quality cost concept.

At the Continental Illinois, 'prevention costs represent 7.5 per cent of quality costs, and 2.8 per cent of total operating expense. Appraisal costs represent 59.3 per cent of quality costs and 22.1 per cent of total operating expense. Internal failure costs represent 13.1 per cent of quality cost and 4.9 per cent of total operating expense. External failure costs represent 20.1 per cent of quality cost and 7.5 per cent of total operating expense.' (Eldridge 1983). Despite spending nearly a quarter of their total operating expense on appraisal, the bank has to spend 7.5 per cent of its costs on correcting errors that the appraisal system has failed to catch before they got into the hands of the customers.

This is a typical state of affairs at an early stage of a quality programme. Enormous effort is spent on inspecting product, and relatively little on preventing the service going wrong in the first place. It is clear also at this stage that 'quality costs', however defined, are a significant proportion of the company's operating expense, quite sufficient to alert top managerial attention. The next thing to do is to reduce the level of these costs, typically by increasing prevention to reduce failure costs (and perhaps in the short run, appraisal). It has been found in numerous studies that a doubling of prevention costs (to say 6 per cent of operating costs) can reduce appraisal and failure costs by a quarter or more, thus reducing the total quality cost.

Another way of addressing the problem of costing the quality of service is the Cost of Error approach. This is a service industry version of the Failure Mode and Criticality Analysis used in manufacturing design to estimate probability of machine failures and to assess safety risks. In this technique, the costs are divided into three categories:

- hard costs, that is, those costs that directly arise from the error, such as rebates, re-working, waste, time and effort spent answering complaints etc.

- soft costs, the costs that arise as a result of error — for instance if a chamber maid spends 15 per cent of her time repairing errors in her room-cleaning routine, somehow that 15 per cent has to be made up.
- opportunity costs, future sales lost as a result of the error.

The last category, which has to be estimated, can be a surprisingly large amount. In 1986 Club Med discovered through market research that 80 per cent of new customers were satisfied with the holiday, and that of these about a third would return. These loyal customers would make an average of three further Club Med holidays. The average new customer was worth a life-time stream of contributions of about $4,000. Thus the new customers of 1986 were likely to be worth some $192 million over the next few years. With the magic of computerized spreadsheets, the company was able to ask some 'what-if' questions, the answers to which made them sit up and think. First of all they reduced the proportion of satisfied customers by 1 per cent to 79 per cent; then, still pursuing the effects of a lower quality standard, they reduced the number of repeat visits to 3.5 over a life-time. The combined effect of these apparently insignificant changes was to create a pre-tax loss of $26.1 million (Heskett et al. 1990, 74)

Cost of error analysis

Annual cost of this type of error = Cumulative expected cost of all consequences of the error × probability of the error occurring × number of events in which the error could occur per year.

1. Define the error: In drawing up and analyzing the service delivery specification, failure points will be identified. These might be, for instance, 'wrong bank statement sent', 'hotel bed not made' 'wrong sandwich filling'.

2. Determine the consequences: As soon as an error is made, a tree of consequences starts. For instance, if the error is spotted before it reaches the customer, it is corrected, if not, the customer may or may not complain, a complaint may or may not be rectifiable, the effort of rectification may be large or small, if rectified the action may or may not mollify the customer.

3. Establish a cost for each consequence: Each of these consequences has a cost; the cost of replacing the wrongly filled sandwich, the waste of the first sandwich, the time spent assessing the problem, the manager's time spent mollifying the customer, etc. A large proportion of service company time is spent checking for human error and these costs are included here. If the checking system fails, there is a strong chance that the customer may not complain, but simply not return, thus taking an expected stream of revenue elsewhere. We have seen that for Club Med, this could cost them £26 million.

4. Establish the likelihood of each consequence: This is easily done for errors discovered by in-house checkers, not so easily for those errors that slip through the net. Most customers do not complain, so the company is usually saved the immediate costs of setting errors right at the expense of a very much reduced likelihood of repeat business. The TARP research makes it clear that customers generally prefer to vote with their feet rather than to make a fuss.

5. Calculate the expected cost (multiply 3. x 4.): The expected cost is a statistical concept. It combines the cost of an event occurring with its likelihood. This allows us to compare the cost of certain errors (food poisoning, wrong bank statements, over-charging) that may be infrequent but critical, with others that were more frequent but less serious (stale flowers in the hotel bedroom, a few days late with a standing order, stock-outs).

6. Calculate the total cost of the error per year: Having identified the expected cost of each consequence, the next step is to estimate its frequency. For instance a 300-room hotel might generate 37,000 check-ins; for each of these there is a definable probability of failure (and we have just calculated the cost of failure). It is therefore an easy matter to calculate the annual cost of this kind of failure.

That is the theory. The problem is how to do it. In practice this means moving away from cost and finance-based figures to activity-based figures. Certainly quality costs and cost of error figures are good ways occasionally to summarize direct quality management activity. It is increasingly clear that customers generally do not care whether you are making a profit or not; they do care about the product. Profit, like happiness, follows from activity, not the other way about. So, measure and develop activities, and profit will follow; measure only costs and all thoughts will be on how to reduce them. (Measures of activity also have the added advantage that they are much less likely to be financially sensitive.) Tom Peters has amply demonstrated, especially in his book *In Search of Excellence* that companies thrive in proportion to senior management's knowledge of both the product and the market.

What to measure: activity-based

Any measuring system is established to enable comparisons to be made. Cost measuring and reporting enables managers to compare the values they get from the company at the end of the year with other values. This gives them a sense that they are broadly on the right track, and also enables them to balance financial calls such as wages, purchase prices, sales prices etc. but for day-to-day management of an activity, most of these matters are irrelevant. Decisions about these matters have to be made from time to time; what requires daily monitoring is activity. The analysis of the service delivery system will identify key revenue generating factors. Following the employee involvement philosophy, the progress made with these factors should not be locked in the managers' desks, but be publicly displayed for all to see. As Tom Peters writes, 'simple visible measures of what's important should mark every square foot of every department of every operation'.

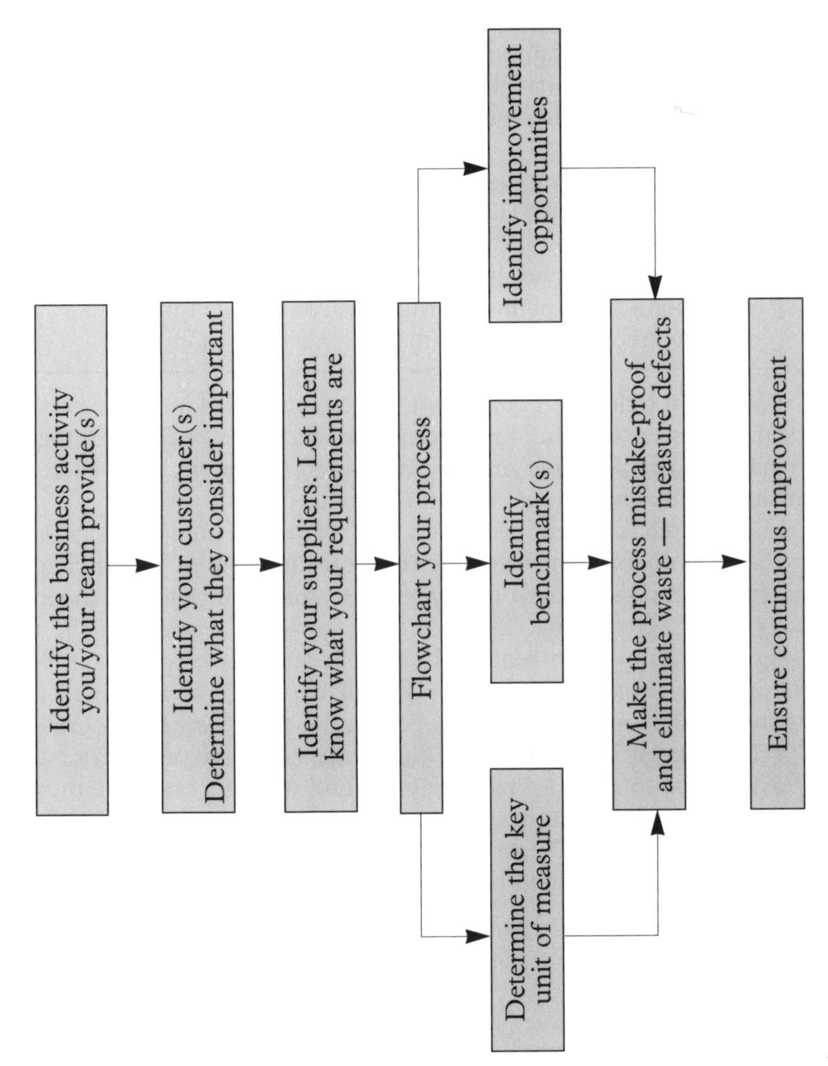

Source: Northern Telecom

SOME TYPICAL SERVICE STANDARDS

- average and range of speed of service fulfilment from entry point to exit.
- average and worst case length of queue.
- number of stock lines not available.
- number of clerical errors in customer communications.
- all telephone calls answered within 4 rings, and all holding callers reverted to every 30 seconds.
- customers acknowledged within 20 seconds of entering the premises.
- customer's name used at least once in the conversation.
- number of times all bedrooms were cleaned before 12.00.
- number of on-time deliveries/arrivals.
- number of service system improvements introduced this quarter.
- customer compliments/criticisms/complaints ratios.
- executive hours spent with customers per month.

The Malcolm Baldrige requirements specify that tracking indicators should represent attributes that link to customer satisfaction as well as to operational performance. In establishing measurements and indicators, as in all aspects of the quality approach, the customer must come first.

Recent research has identified a new way for service companies in particular to look at customers. Instead of focusing on the costs of a particular transaction, this method sees each customer as a stream of potential income over time. As loyalty grows, the customer tends to spend more, and the costs of handling each transaction reduce. Reichheld and Sasser give examples of a credit card operation, which expects to make a loss of $51 on the first year, and then profits rising every year to $55 in the fifth year; a car-service company which calculated to make more than three times the profit from a fourth-year customer than from a first-year one. The strategic objective therefore is to hold on to existing customers: manufacturers look for zero defects, service companies look for zero defections (Reichheld and Sasser 1990).

The reasons for the added profitability were:

- reduction in the extremely high cost of acquiring new customers. This cost can be as much as five times that of retaining existing customers.
- long-term customers tend to buy more.
- long-term customers tend to place frequent, consistent orders and are therefore less costly to service.
- long-term customers refer potential new clients.
- retaining customers makes share growth difficult for competitors.
- satisfied customers are often willing to pay premium prices for service they know and trust.

Even by reducing customer defections by as little as 5 per cent boosted profits in a range of service industries from 20 per cent to 85 per cent. For instance, a 5 per cent increase in customer retention in mail order companies drove profits up 20 per cent; in software the same increase in customer retention gave a 26 per cent increase; in insurance brokerage 50 per cent. Numerate thinking of this sort encourages companies to put a very high value on the 'moment of truth'.

The US management consultancy Bain has explored this concept. It has found and developed a simple technique for measuring customer retention. The customer retention rate is defined as the percentage of customers at the beginning of the year that are still there at the end. As a firm's retention rate rises, so does profitability and the longevity of the average account. Bain suggests that an 80 per cent retention rate means a five-year account length, and a 90 per cent retention a ten-year 'life'.

As with all service company measurements, it may be necessary to devise creative ways of ascertaining defections or repeat customer percentages. As Reichheld and Sasser put it, 'for some businesses, the task of spotting defectors is challenging even if they are well defined, because customers tend to be faceless and nameless to management. Businesses like retailing have to find creative ways to "know" their customers.' One company cited gives its customers a free discount card, so that all subsequent transactions can be traced; a restaurant in Maryland tracks its repeat customers through the booking log; we have already referred to the hotel in Hong Kong which uses passport information and a database to check return frequency.

However it is done, the company must accept that the customers (and their friends) represent a life-time stream of business, not just a single transaction. A small change in this stream can have dramatic effects on the profitability of the business. No other measure will so quickly and sensitively detect that, for instance, although your level of service may be even better than it was last year, it has been overtaken by the rivals down the road. It is therefore essential that a sensitive finger be kept on this pulse. The defections or repeat business ratios should be carefully monitored as the first indicator of service quality health.

Monitoring the stream of defections is undoubtedly a valuable indicator of overall performance. However it does not indicate in itself which aspects of the quality service are proving successful and which not. This requires further analysis. The organizers of the PIMS database break quality into two aspects:

- perceived quality.
- conformance quality.

Perceived quality is in effect the quality of a product as perceived by the customer, and conformance quality is the ability to reproduce that perceived quality over and over again. Perceived quality is the value the customer gets from the moment of truth, a superbly cooked meal, or a piece of timely helpfulness; conformance quality is the value that the company delivers from being assured that the same meal or the same helpfulness is not simply the inspiration of the moment, but a regular part of the firm's activity.

The two forms of quality performance deliver different values to the profit and loss account. As the PIMS database specialists put it, 'superior perceived quality gives your business three options, all of them good.

First, you can charge a higher price for your superior quality offering and let the premium fall right to the bottom line.

Second, you can charge a higher price and invest the premium in R & D and in new products to ensure your perceived quality and market share in the future.

Third, you can offer better value by charging the same price as your competitors, but for your superior product/service offering. This will allow you to build for the future by gaining market share. The gain in market share means volume growth, rising capacity utilization, and ultimately capacity expansion allowing you to introduce new equipment that embodies the latest cost-saving technology.' We have already seen the importance of market share in achieving a high return on investment.

Achieving superior conformance quality yields two key benefits. 'First, it means a lower cost of quality than that of competitors, and thereby a lower overall cost. Second, conformance quality is one of the key attributes that count in purchase decisions. So achieving superior conformance quality yields both lower cost and superior perceived quality — a double benefit.' (Buzzell and Gale 1987, 104–7). The difference can be compared to the difference between accuracy (perceived quality) and precision (conformance quality).

In target shooting, accuracy implies nearness to the centre of the target; precision on the other hand means the nearness of one shot to another. A gifted amateur shot is likely to be accurate, but imprecise, his shots clustering round the centre of the target more or less imprecisely. A professional shot values precision, a tightly bunched group of hits, first. The theory is that if one has sufficient control to shoot repeatedly into the same tight cluster, it will be a relatively easy matter to relocate the cluster to the centre of the target. The same applies to quality management. The gifted amateur can often provide excellent service; only a professionally managed and controlled organization can deliver that service with unremitting precision, time after time. This is the true meaning of quality.

In establishing measurement systems, therefore, the logical place to start is to concentrate on assessing and then controlling sources of variation, thus increasing conformance quality. Luckily conformance is easier to assess than perceived quality.

In Chapter 4 we discussed the five key service quality factors identified by Zeithaml, Parasuraman and Berry in the United States. These factors obviously provide a starting point for a strategy of measurement. The factors identified were:

- reliability.
- responsiveness.

- assurance.
- empathy.
- tangibles.

The original research was based on ten factors: tangibles, reliability, responsiveness, competence, courtesy, credibility, security, access, communication, understanding/knowing the customer. From the responses to questions on customers' expectations and perceptions of service delivery, they found that the first three factors stood out clearly, but the remainder tended to overlap in customers' perceptions.

As a result they combined competence, courtesy, credibility and security into a new factor called assurance; and access, communications and understanding into a new factor called empathy. In using these results to establish a Key Result Area measurement system, businesses may find that 'empathy', with its combination of the practical (access and communication, which could as well be classed with tangibles) with the personal, a little difficult to handle. There is also something to be said for adding the ability to recover from errors to this fundamental list.

Zeithaml, Parasuraman and Berry found that customers looked for different weights of these factors from different service organizations. For instance, 25 per cent of banking customers looked for empathy, whereas only 10 per cent mentioned needed this from long-distance telephone services. In every case however, by far the most prominent requirement was reliability. Luckily this is relatively easy to measure (Zeithaml, Parasuraman and Berry 1990).

Reliability is defined as the ability to perform the promised service dependably and accurately. Customers enter the business with a set of expectations; if these are met, all is well, if not, disappointment occurs. Reliability measures therefore start with these expectations, for instance:

- how often was the burger served in more than 30 seconds?
- how many deliveries were made at the time promised?
- how accurate was the estimate of price?
- how often did follow-up letters revise earlier statements?

The second factor identified in market research was responsiveness. Once again this can be reduced to a series of time measures. Speed of response to telephones is often measured, as is speed of response to breakdowns, or service calls. Responsiveness is allied to assurance, the third factor identified. This is the 'safety' factor in a company's service. In days gone by, professional firms made much of this. If you went to the reliable old firm of X and Partners, the letters might not always be answered, the office might be a bit shabby, and the empathy might be weak, but you could trust them not to run off with your life's savings.

With this factor, as with the remaining two of tangibles (appearance of the firm) and empathy (caring individualized attention), it is hard to generalize about how to measure them. Different businesses have different ways of expressing these values. However it is clear that they are part of the package that customers require from service companies. It is therefore important to analyze how they are being delivered, and to establish indicators to measure them.

The creation of an effective measurement strategy can demand considerable creative ingenuity. In some areas it may be possible to use the techniques of benchmarking, or learning from other companies' experience. The theory behind this idea, which is an essential part of the Malcolm Baldrige protocol, is that excellent companies develop special expertises. There are only a certain number of operational areas, so fruitful exchange can often occur between non-competing companies. For instance supermarkets have extremely tightly organized stock control systems; stock control is a common problem in many other industries. Clearly a study of how supermarkets do it could be extremely stimulating. Equally, book and record stores have common industrial problems; no doubt each could learn much from each other, and from an industry such as pharmaceuticals.

It is frequently said that service quality deals in such intangible elements that measurement is not possible: this half-truth simply lets intellectually lazy managers off the hook. There are certainly some things that cannot be measured directly, just as physicists cannot measure gravity: they can however measure the effects of gravity. Much of the history of physics is tied up in the history of the development of more and more sophisticated measuring devices: perhaps the history of quality management should be written in the same terms.

ACTION

- **what is the cost of error or quality cost in your business?**
- **what proportion of customers repeat?**
- **what is the life-time value of each type of customer?**
- **what are the relative weights of the five key factors of service?**
- **should some of the factors be subdivided for your business?**

SUMMARY

Measurement is the most important technique in developing Total Service Quality. Key result areas must be defined and results displayed prominently to relevant workers. This chapter identified categories of measurement.

- *financial, especially quality costs, and cost of error studies.*
- *activity, overall, the most important measure is the customer defection rate. Internally distinguish between measurements of precision (conformance quality) and accuracy (perceived value), and for both elements on the five key values of service quality: reliability, responsiveness, assurance, empathy and tangibles.*

CHAPTER 14

Implementing Total Service Quality

'The best way to predict the future is to invent it.'
Alan May

Perhaps you do not have a problem. Perhaps you see the introduction of Total Service Quality management as a technical question of writing down the best practices into a set of documentation, consulting IQA and pinning your Q-Mark registration on the wall. If you do, then you should consider these questions:

1. Does your organization have a USP or mission statement (see Chapter 3)? Is it known and lived by all the employees? Are all employees working to the same policy, standards and priorities? How do you know?
2. Have you identified your customers' key needs, and do you measure how far you are meeting them?
3. In what ways is your process better or worse than your main rivals'?
4. What are you doing to instil pride of workmanship in your employees?
5. How many employees had training for five full days last year?
6. What precise impact did the last three system improvements introduced have on your throughput?

Most Irish managers would draw blank answers to these questions. Yet they, not written procedures or ISO registrations, are at the heart of the service quality quest. Get them right, and everything else will follow.

The journey to Total Service Quality is long. If the requisite change of corporate and personal culture is to be achieved, long lasting commitment will be required from the top management in fair weather and in foul. Like marriage, it is not to be undertaken lightly or ill-advisedly.

If mishandled, it could shake the company's long-established foundations, without achieving any serious results. Nothing serious will be done in less than a matter of years. A small new firm might achieve wonders in only one or two years; a large old firm, with well-established ways, no apparent threats, many conservatively minded employees and a prickly union/management relationship, will take considerably longer. Given interest and enthusiasm from top management in the Total Quality road, one question remains: what exactly must I do to guide the firm to the goal?

The first approach must be to examine closely the guidelines to service quality management laid down by quality standard and quality award organizations. In Ireland there are four protocols to consider:

1. The Irish Quality Association's Service Quality mark.
2. The international standard ISO 9000.
3. The Malcolm Baldrige National Quality Award Guidelines.
4. The European Quality Award.

Service Quality Mark: this is awarded after the company has completed a questionnaire describing its quality system, and the system has been inspected by an auditor from the IQA. The Auditor-General of the IQA will usually visit the company before it completes the questionnaire to discuss with management the requirements of the Approved Quality System.

During the inspection the auditor will explore how far the company's quality management system assures delivery of the promised standards of performance. Four areas of management are particularly examined:

- the management of quality performance: this includes such questions as:
  - does top management participate in setting and monitoring quality activity?
  - do the stated quality objectives relate to all the key aspects of the service?
- the quality system itself.
  - are individual responsibilities for quality clear?
  - are adequate human and material resources available to deliver the quality objectives?
  - what internal and external auditing of the quality systems take place?
  - how are quality results and difficulties reported to management?
- operational elements.
  - are customer needs clearly identified?
  - are quality control points during delivery of service specified?
  - if a failure in service takes place, what system is in place to prevent it happening again?
- support elements.
  - is there a staff training plan?
  - how are quality requirements communicated to staff?

ISO 9000: the ISO Quality Management and Quality System elements — Part 2: Guidelines for Services (9004–2) has identified four key aspects of the quality system which should be considered when drawing up the quality road-map. These 'key aspects' are:

1. **Management responsibility:** This includes establishing a policy for service quality and customer satisfaction. Management is also responsible for ensuring that the service providers are committed to the policy, and that a philosophy of continuous improvement is actively pursued.

2. **Personnel and material resources:** Management must ensure that the skilled people and the appropriate equipment are available to meet the policy targets. This means paying particular attention to recruitment, training, motivation and reward systems.

3. **Quality system structure:** The quality system is defined and documented (at least in outline) in the quality manual, which should contain:

- a system-wide quality policy.
- the company's quality objectives.
- the structure of the organization.
- a description of the quality system.
- the quality practices of the organization.
- the structure and distribution of the quality system documentation.

For each individual service, there should be a further set of documents, consisting of:

- service brief.
- service specification.
- service delivery specification.
- quality control specification.

The quality control specification consists firstly of a quality plan for the particular service outlining the objectives of the quality systems as such. This is a combination of service brief and service specification for the quality system. Secondly there should be a set of quality procedures backed up by a set of quality records.

4. **Interface with customers:** Management must ensure that effective channels of communication are established between the organization and its customers. This involves both listening to the customers and keeping them informed. Difficulties in communication or interactions with customers, including internal customers, should be given prompt attention.

Finally, the ISO 9004–2 Guidelines list four generic operating areas that exist one way or another in all service operations: marketing, service design, service delivery and performance appraisal and improvement. In order to judge how far the documented quality system has progressed, it would be desirable to be able to present the specifications and procedures as the outcome of a grid or matrix. Thus the specification would detail the personnel and material resources required in marketing, service design, service delivery and performance appraisal and improvement.

A more sophisticated Y-shaped matrix could combine the strategic guidelines, the key aspects and the operating areas into a three-dimensional display. This would present a comprehensive index and control chart for the whole quality journey. It should be said that the use of Y-shaped matrix charts is not widespread, and they are therefore regarded as difficult to interpret.

Malcolm Baldrige National Quality Award: this award was initiated in 1988 as the American version of the long established Deming Award in Japan. As an award scheme, rather than a registration scheme like the Q-Mark and the ISO 9000 standard, it issues awards to only six companies a year. The protocol for the award however makes a detailed and formidable checklist for the assessment of any company.

The key concepts on which the scheme is built constitute a checklist of modern quality thinking. They are:

- quality is defined by the customer.
- the senior leadership of the firm needs to create clear quality values and build the values into the way the company operates.
- quality excellence derives from well-designed and well-executed systems and processes.
- continuous improvement must be part of the management of all systems and processes.
- companies need to develop goals, as well as strategic and operational plans to achieve quality leadership.
- shortening the response time of all operations and processes in the company needs to be part of the quality improvement effort.
- operations and decisions of the company must be based on facts and data.
- all employees must be suitably trained, developed and involved in quality activities.
- design quality and defect and error prevention should be major elements of the quality system.
- companies need to communicate quality requirements to suppliers and also work to elevate supplier quality performance.

The Baldrige Award is scored out of 1,000 points, broken down into seven categories, as follows:

1. Customer satisfaction — 300 points.
2. Quality results — 180 points.
3. Human resource utilization — 150 points.
4. Quality assurance of products and services — 140 points.
5. Leadership — 100 points.

6. Information and analysis — 70 points.
7. Strategic quality planning — 60 points.

A copy of the full current protocol of the Malcolm Baldrige Awards can be obtained (free) from the United States Department of Commerce, National Institute of Standards and Technology, Gaithersburg, Maryland 20899, USA.

European Quality Award: this award, which was launched in October 1991, is very similar to the Malcolm Baldrige award. It was established to identify model companies excelling in the late twentieth century, irrespective of size or business. The scheme is for companies and subsidiaries based in Western Europe. Subsidiaries of non-European companies must act independently of the head company.

The award protocol breaks down the points by percentages as follows:

1. Customer satisfaction: the perceptions of external customers, direct and indirect, of the company and of its products and services (20%).
2. People: the management of the company's people and the employees' feeling about the company (18%).
3. Business results: the company's achievement in relation to its planned business performance (15%).
4. Process: the management of all the value-adding activities within the company (14%).
5. Leadership: the behaviour of all managers in transforming the company towards Total Quality (10%).
6. Resources: the management, utilization and preservation of financial, information and technical resources (9%).
7. Policy and strategy: the company's vision, values and direction and the ways in which it achieves them (8%).
8. Impact on society: the perceptions of the community at large of the company. Views on the company's approach to quality of life, to the environment and to the need for preservation of global resources are included (6%).

Fuller details of the European Award can be obtained in Ireland from the offices of the Irish Quality Association.

The TQM gurus

The protocols of the various registration and award organizations give valuable pointers to matters to be attended to on the journey, and where attention should be directed. They have hardly provided the action plan that is the answer to the CEO's question: what exactly must I do to implement the ideas of Total Service Quality?

Like every explorer seeking a way through uncharted territory, case studies of the approaches taken by other companies as published in magazines and journals relating to quality can be extremely helpful. These can be found in the libraries of the Irish Quality Association and the Irish Management Institute. The case studies tend (not surprisingly) to report only success stories. It would be instructive also to have reports and discussions of initiatives that failed.

Another tack would be to read what the numerous experts on the subject have to say. Every writer on Total Quality has a slightly different answer to the question. Three of the best known are American, and have drawn up progressive steps or pointers to quality management.

Dr J. Juran is one of the two grand old men of quality who took the best American practices to Japan after the war. The success of Japanese industry in later years has been largely ascribed to the insights delivered by Juran and W. Edwards Deming which were accepted and acted on by Japanese industry. He has written many books, and is founder president of the Juran Institute.

In his view less than 20 per cent of quality problems are due to workers. He therefore stresses the importance of management commitment in the development of a Total Quality strategy. He believes this management commitment is most effectively achieved by a good quality cost reporting programme. Top management, he says, speaks only the language of money. Therefore to appeal to them, concepts must be explained in those terms. He identifies ten steps to quality improvement:

1. Build awareness of the need and opportunity for improvement.
2. Set goals.
3. Organize to reach the goals
    - set up a quality council.
    - identify problems.
    - select projects.
    - appoint teams.
    - designate facilitators.
4. Provide training.
5. Carry out projects to solve problems.
6. Report progress.
7. Give recognition.
8. Communicate results.
9. Keep the score.
10. Make annual improvement part of the regular processes of the company.

Books: *Quality Control Handbook*, *Management Breakthrough.*

Dr W. Edwards Deming was the other of the two famous gurus who helped Japanese companies after the war. Such was his importance there that the annual quality prize in Japan is named after him. Deming is by training a statistician, who defines quality as a 'predictable degree of uniformity and dependability, at low cost and suited to the market'. Deming starts from the idea that quality has three dimensions (expressed in service terms):

- quality of design.
- quality of service delivery.
- quality in the performance/expectation gap.

As the quality in the expectation/performance gap changes, the design has to be improved, and so a cycle of continuous responsiveness to the market is established.

Deming defines variation from the ideal as the basic cause of loss of quality. This variation has two causes. The first is common variation, which is built into the process; the second is special or assignable variation which is caused by a change in the system. He claims that much management time is wasted by a confusion between the two. Common variation from the customer's expectations might be caused by systematic factors such as poor lighting, poor training, lack of suitable materials; special variation might be caused by a new operator, a poorly explained new system or broken equipment. Clearly systematic or common variation factors are much more difficult to eradicate than assignable variations.

To help management to address the problems of process control and reduction of variation, Deming has established his famous 14 points. Since management is responsible for 85 per cent of the problems, most of these points are addressed directly at them:

1. Constancy of purpose: think long term; nothing as serious as this can be done in three months.
2. Adopt the new philosophy of never-ending improvement: the key motivating idea in Deming's philosophy is that we are living in an age of change. 'Ah sure it'll do' is no longer good enough (if it ever was!). The implications of making improvement the key activity are profound.
3. Cease dependence on inspection: many service companies such as banks rely on inspection checks to ensure that the product is right; Deming says that this is shutting the stable door after the horse has bolted. Concentration should rather be put on getting it right first time, and designing systems that ensure this.
4. Change the way you choose suppliers: the total cost of purchases is not just the price on the invoice. It includes the cost of specifying, ordering, arranging delivery, screening, rejecting, re-working, re-ordering and so on. To eliminate this cost, move towards a much closer long-term relationship with suppliers. The effort put into this relationship dictates that there would be fewer of them.
5. Constantly work to improve the system: everything an organization does is capable of improvement. It is management's responsibility to drive the improvement process. This means constantly identifying problems, collecting data, analyzing the data, evolving solutions, testing the solution, establishing the solution, then identifying the next problem.
6. Establish training for everyone: training programmes should be set up to cover organizational goals, new employees' orientation, supervisory development, management skills, team building, problem-solving techniques, and new techniques.

7. Develop local leadership: empower the supervisors and workers to take responsibility for providing solutions to their own problems. This particularly applies to supervisors, whose model must be as a coach and leader, not a whip hand on a slave plantation.
8. Drive out fear: our objective is to stimulate the creative enthusiasm of the workers. As Maslow's hierarchy makes clear, these aspects of every human can only be addressed once fear for the lower needs is removed. Lack of job security, aggressive performance appraisal leading to loss of status, lack of understanding of company goals, being blamed for errors caused by bad inputs: all these create an atmosphere of fear and so of paralysis.
9. Break down barriers between departments: barriers inside the organization mean a specialized division of labour and a heightened sense of teamwork, which are good. They also mean ignorance of the whole corporate picture, which results in errors and confusion.
10. Eliminate slogans and meaningless targets: 'Do it Right First Time', 'Cut costs Now', 'Use the Goggles Provided', 'Increase productivity by 18 per cent': as the rulers in the Soviet bloc have found, these exhortations are no substitute for providing the conditions by which the target is to be achieved.
11. Concentrate on causes, not quotas: work standards, bonus schemes and quotas are frequently used to stimulate worker performance. Unfortunately they nearly always result in quantity being produced before quality.
12. Stimulate pride in workmanship: people like to be proud of their work. As the quotation in Chapter 5 from Studs Terkel exemplifies, this potential is often ignored by managers, to the serious detriment of the quality and productivity of the workforce as a whole.
13. Develop self-improvement: Claus Møller says that a workforce with high self-esteem will produce good work. This self-esteem is of one piece, both inside and outside the work-place. Develop self-esteem by encouraging self-development: broaden horizons, stimulate, push people to build on their interests.
14. Do it now: create a structure which will push the prior 13 points every day.

Books: *Quality, Productivity and Competitive Position*, *Out of the Crisis* (1986), see also H.S. and S.J. Gitlow *The Deming Guide to Quality and Competitive Position* (1987).

Philip Crosby is widely known as the author of *Quality is Free* and other books advocating 'zero defects' concepts. This is perceived not as a way of harassing workers, but as a standard for management to pursue. His core idea is that services conform to requirements or they do not; the target must be zero non-conformances, and the method by which management achieves this is prevention. He suggests a 14-step path to quality improvement:

1. Establish personal management commitment to quality.
2. Bring together representatives from each department to form quality improvement teams.
3. Identify by numerical values the state of quality throughout the company.
4. Evaluate the cost of quality and explain its use as a management tool.
5. Raise the quality awareness of all employees.
6. Encourage the discussion and solution of quality problems by corrective actions.
7. Establish an ad hoc committee for the zero defects programme.
8. Train supervisors to actively carry out their part of the quality improvement programme.
9. Hold a 'zero defects' day to let all employees realize that there has been a change.
10. Get supervisors and workers to establish improvement goals for themselves and their groups.
11. Remove the identified causes of error.
12. Recognize and appreciate those who participate.
13. Establish quality councils to communicate on a regular basis.
14. Do it all over again to emphasize that the quality improvement programme never ends.

The Crosby consultancy organization, which has an office in Britain, offers company-wide training programmes based on these principles. A number of Irish companies, particularly in the computer field, have successfully applied the principles.

Books: *Quality is Free*, *Quality Without Tears*, *Running Things.*

John Oakland is perhaps the most appropriate for Irish companies. He has written a particularly straightforward introduction to Total Quality Management, which although primarily aimed at manufacturing concerns has much to say that is relevant to service companies. He does not assume, as many authors do, a very well established quality structure. His road map is therefore particularly suitable for service companies that may not have progressed very far down the formal quality road. Oakland's ideas are also more on the Irish scale. Unlike many American gurus (and the Malcolm Baldrige scheme) he does not think of a 'small company' as one with only 400 employees!

He identifies twelve steps on the path:

1. Understanding: the hardest step for action-oriented managers. Do not plunge the company into the self-assessment and change required for Total Service Quality unless you understand exactly what you are doing and why. One video or one book is unlikely to give you sufficient information or understanding for so serious a step. Spend as much time on knowing what Total Service Quality means as you would on a major investment.

2. Commitment and policy: the understanding must be expressed by commitment, which in turn is expressed by policies plans and actions for Total Quality.
3. Organization: making this happen requires the appropriate organization, whether from inside or outside (in the form of a consultant) or both.
4. Measurement (especially costs): collect real information on how the company works, in cost and activity terms. Establish new measures that relate to customer needs, not organizational convenience.
5. Planning: bring in managers to plan the improvement process. Their commitment to the whole process will be critical to its success.
6. Design: use the systems now established to examine the service brief and the service design. Does it really meet the customers' needs? Can you prove it?
7. Systems: If you have not done so before, begin to write down the formal quality system. Start from the procedures and work upwards, not from the manual downwards.
8. Capability: establish monitors to ensure that the process meets the requirements both on average and in times of pressure and stress. Does it struggle to deliver standards at times? If so, what can you do about it?
9. Control: the quality system will include modules relating to the inspection, checking and testing of the output. Make sure that this is giving the results, that the results are recorded, that the records are analyzed, and that the analysis is used for the continuous improvement process.
10. Teamwork: now is the time to communicate more deeply the quality and problem-solving objectives of the programme, and to establish quality management teams that cross functional boundaries to address system-wide problems.
11. Training: to achieve the best results with step 10 above, a full scale training programme, first covering management and then service workers needs to be established. This should cover the quality skills as well as the departmental skills required to deliver the customers' requirements.
12. Implementation: Steps 1 to 11 should give you an effective working quality system more than adequate to gain the Irish Quality Association's Quality Mark and registration under ISO 9000. The next step is to develop the Total Quality approach that addresses the more sophisticated requirements of the European Quality Award or the Malcolm Baldrige Award.

Books: *Total Quality Management*, *Quality Improvement Through Standards.*

Six strategic considerations

Every service organization is different. They have different starting points, different resources, different structures: it is therefore not sensible to declare one route to Total Service Quality as always superior to another. Many companies have followed the paths laid down by the gurus with evident success;

CASE STUDY: DEVELOPING A QUALITY CULTURE

The Quality Focus initiative in Aer Lingus is essentially about driving a culture change. As you know organisation models span a spectrum starting with a control model of organisation and finishing with a commitment model of organisation.

The former is an organisation with strict hierarchical lines of management, a corporate centre that is generally out of touch and insulated from the customer; middle management in very much a control or blocking mode and to quote Tom Peters 'a thick impermeable membrane separates the customers from the organisation with specifically designed receptacles for measuring customer service'.

The latter type of organisation is based on the concept of eliciting commitment whereby the whole thrust is towards a focus on meeting customer requirements, continuous listening to changing customer requirements and middle and senior management acting as facilitators or coaches to staff, to encourage and facilitate staff in meeting changing customer requirements.

Essentially this is the type of culture we are aiming for with Quality Focus. In other words, we are trying to move from a fairly traditional western European type of organisation to the Japanese models of organisation as described in theory Z.

Source: Aer Lingus

less has been heard of the failures. It is often sensible to appoint a consultant, though this should never be a substitute for understanding and choice on the part of senior management. Since each consultant brings his or her preferred solutions or approaches to the problem, choosing a consultant is equivalent to choosing a style of solution.

At the beginning of the journey, the company must identify where it is going, and a set of measurements of progress. A plan should be developed, and monitor or control points established. The six key strategies discussed in this book are suggested as the guiding lights by which Total Service Quality will be achieved. The more satisfactory the performance in these six areas, the nearer is the goal. They are:

1. Client focus: A company lives or dies by what it does for the customers. No one patronizes a firm because it has a good social club or pension scheme for the employees. Everything done on the premises should be relatable to the customers' needs. This was discussed in Chapter 3.

2. Employee involvement: The quality of service is peculiarly dependent on personal relations between the service giver and the customer. The more the employee is involved in the service planning, and the more he or she is committed to the objectives of the firm, the better the service. The USP/mission statement is important here. However, the key to employee involvement is in the real culture of the firm as evidenced not by words but by acts. A firm that proclaims the virtues of consultation and retains executive toilets; a firm that announces the importance of the customers, and reserves the best spaces in the car park for top management; a firm that urges loyalty and regulates pay by status rather than contribution: all these are hypocritically saying one thing and meaning another. This was discussed in Chapter 4.

3. Knowledge of the process: There is no substitute for a full and detailed knowledge of the process. The more detailed it is, the better. This means intense focus on the inputs, the bottlenecks, the activity flows, and the contributions of each sub-process to the whole. Management must know exactly what the problems and difficulties are at each stage, preferably from personal hands-on experience. Improvements can only come from this full knowledge.

4. Measurement: Counting and measurement of key variables is the single most important technique in the quality armoury. In Chapter 13 we identified quality cost and zero defections as critical indicators of corporate health. There should also be measures of reliability, responsiveness and the tangibles of the firm. Finally the philosophy of constant improvement suggests that we should measure these also. A company that takes improvement seriously can, as we suggest, make this a serious focus of management activity.

5. Quality systems: Documented quality or operational systems are the formal foundation on which the high aspirations of Total Service Quality are hung. The meticulous detail and control implied by these systems, especially the ISO 9000 series, are often time consuming and may sometimes seem out of balance with the needs of the firm and its customers, but they are the essential basis of a good quality system. Just as measurement without analysis and action is useless, so is an elaborately documented and maintained quality system without a good product. The quality system is designed to maintain and control the level of quality you have obtained. It is not in itself a guarantee that you have attained a level that will satisfy the customers.

6. Continuous improvement: This is the heart of the Total Service Quality approach. We are not living in a stable world where the successful disciplines and ideas of yesterday will suffice. Change is a constant. Therefore it is not enough to be as good as we were yesterday; we must be different in response to different conditions. The five previous strategic guidelines are sterile without continuous improvement, which should form the basis of every management approach. Like everything else, improvements should be counted.

What not to do in Introducing Total Service Quality

1. Don't move too quickly. Let the ideas grow.
2. Don't fail to communicate everything to everyone, before you do anything, not afterwards.
3. Don't let short-term urgencies defeat the long-term objectives.
4. Don't become obsessed with ever increasing standards for their own sake.
5. Don't use threats to force improvement.
6. Don't assume that service will improve just because the boss says it should.
7. Don't rely on slogans, pep-talks and posters.
8. Don't assume that middle managers will automatically accept the improvement process.
9. Don't let the initiative fail for lack of resources.
10. Don't assume that equipment or training is adequate.

In Chapter 2 the techniques of understanding customers' needs and expectations were discussed. In doing this, excellent companies establish a real relationship between the supplier and the receiver of a service. In practice, however, this happens too rarely. Consumers do not usually *like* the organizations from which they buy goods and services. They tolerate them; they put up with them. Many businesses are served by loyal employees, who commit themselves whole-heartedly to the firm. Very few businesses succeed in generating similar affection from their customers. How many stout drinkers feel any interest in Guinness as a firm? How many weekly grocery shoppers would not instantly desert Superquinn for Quinnsworth, or vice versa, as soon as a larger or more convenient store was opened? How many motorists care whether they put Esso or Texaco fuels in their cars?

If your company continues to meet the customers' needs better than any other currently available, it will survive. As soon as someone else does it better, your company is in trouble. No matter how long established and famous your firm is, it is what you can do now that matters. A glance at a trade directory of only 10 or 20 years ago will show how in market after market this can be shown to have happened. The restaurant business is a graphic example of this: a certain place becomes popular for a while, and then another takes over. Fascias and sites change all the time, a constant reminder of the volatility of public taste.

Total Service Quality management is a controlled and organized way of ensuring that customers' needs are anticipated and met. It is a blueprint of activities and a management handbook. It encompasses the whole firm, from the accountant to the service providers, from the cleaning staff to the Chief Executive. It is both a philosophy and an action plan. In the highly competitive world of the 1990s, it is a key battlefield. Companies with superb service will do well, others will not. As Dr Deming said: 'You don't have to do this, survival isn't compulsory.'

ACTION

- **establish a senior executive with full functional responsibility for all aspects of customer service.**
- **establish with him/her a customer service improvement strategy.**
- **carry out customer service surveys, round tables, open days.**
- **train all staff in personal customer contact skills.**
- **give top priority to complaints.**
- **ensure that adequate resources are made available.**
- **constantly ask: how can we do it better?**

SUMMARY

If your company continues to meet customers' needs better than your rivals, it will survive; if not, it won't. Total Service Quality is a strategy for ensuring that it does. Above all, it requires commitment in terms of enthusiasm and resources from the top. After that it requires concentration on the six key areas of strategic focus: client focus, employee involvement, knowledge of the process, measuring what's important, documented quality systems, and a continuous improvement mentality.

Appendix: Sample Questionnaires for use in Service Environments

CHECK YOUR ORGANISATION

Does your organisation follow the old principles of control, command, and compliance or the new principles of commitment, consensus, and creativity? Here is a quick test to determine which are dominant in your organisation:

Quality Cultural Characteristics

Check those that most closely describe your organisation.

Old principles in action	New principles in action
☐ Direction driven	☐ Purpose driven
☐ Workers want only pay and time off	☐ Workers want challenge and satisfaction
☐ Many layers of management and slow bureaucracy	☐ Few layers of management and flat structure
☐ Jobs are broken into pieces	☐ Job functions are combined
☐ Work as individuals	☐ Work as teams
☐ Decisions are made at top	☐ Decisions are made at all levels
☐ Worker is replaceable	☐ Worker is valuable and constantly trained
☐ Pay is geared to job class	☐ Pay is linked to skills acquired
☐ Others inspect for quality work	☐ Build in quality and inspect own work
☐ Work to catch mistakes	☐ Work to celebrate success
☐ Divide thinking and doing between managers and workers	☐ Everyone is a thinker
☐ Information is funnelled to the top for decision	☐ All levels collect and use information managers and workers
________ **Totals**	________

Source: D'Egidio 1990

SERVICE QUALITY SENSITIVITY QUESTIONNAIRE

Please tick in appropriate box	Yes	No
1. Has the service strategy been clearly defined, particularly in terms of benefits for the client?	☐	☐
2. The service strategy is communicated in a highly effective manner:		
• externally (intangible aspects become tangible).	☐	☐
• internally.	☐	☐
3. How you calculated the number of moments of truth in your company?	☐	☐
4. Are complaints handled with great care and attention?	☐	☐
Are responses made in writing?	☐	☐
Are follow-up telephone calls made?	☐	☐
5. Have quality standards for every aspect of service been carefully defined?	☐	☐
Were employees involved in defining standards?	☐	☐
6. Were the standards communicated to all employees?	☐	☐
7. Are there intensive training programmes that focus on human resources development?	☐	☐
8. Do employees participate in quality improvement programmes?	☐	☐
9. Are there incentives and rewards for those with superior performance?	☐	☐
10. Is client satisfaction measured regularly by using a representative sample?	☐	☐

Source: D'Egidio 1990

CUSTOMER SERVICE ON THE TELEPHONE

Please tick in appropriate column	nil	little	average	much	very much
1. What confidence do I show to callers?					
2. What efficiency do I demonstrate?					
3. What reliability do I offer?					
4. What degree of caring do I give?					
5. What personal interest do I show?					
6. What length of time, on average, are callers kept waiting before I answer?					
7. What delays occur during calls?					
8. What is the proportion of callers with whom I get angry?					
9. What is the proportion of callers who get angry with me?					
10. What is the proportion of callers who are rude to me?					
11. What is the proportion of callers who are difficult customers?					
12. What is the proportion of callers who are unreasonable?					
13. What improvements could be made in company procedure for dealing with customer enquiries?					
14. What contribution does my work make to establishing good PR?					

Source: Katz 1987

CUSTOMER EDUCATION

Customer education checklist

Place tick in appropriate box

	Yes	No
• Are product usage/assembly instructions clearly and precisely given?	☐	☐
• Is customer understanding of instructions monitored regularly?	☐	☐
• Is customer training freely available?	☐	☐
• Do potential customers know the extent of service back-up available?	☐	☐
• Do PR and media activities inform customers of all product benefits?	☐	☐
• Do the public at large know of the company and its products?	☐	☐
• Do the public think that the company cares?	☐	☐

If the answer to each question is Yes, OK. If No, do something about it.

Source: Katz 1987

ANONYMOUS SERVICE QUALITY QUESTIONNAIRE

THIS WILL TAKE LESS THAN 5 MINUTES OF YOUR TIME

An opportunity to tell us – anonymously – about the service your branch gives you, what we do well, and what we could do better.

QUESTION 1 Sex:

Male ☐ 1 Female ☐ 2

QUESTION 2 Age:

Under 18 ☐ 1 | 18–24 ☐ 2 | 25–34 ☐ 3 | 35–49 ☐ 4 | 50–59 ☐ 5 | 60 + ☐ 6

QUESTION 3 How often do you visit your branch? (That is the branch where you have your main account).

At least once a week ☐ 1 At least once a month ☐ 2 Less often ☐ 3

QUESTION 4 We would like you to rate your branch under a number of headings. Please indicate your rating by ticking the appropriate box opposite each statement on the scale below.

	VERY POOR	*POOR*	*AVERAGE*	*SATIS-FACTORY*	*EXCELLENT*		
General atmosphere of Branch.	☐	☐	☐	☐	☐	1	☐
Being addressed by name when you deal with the Branch.	☐	☐	☐	☐	☐	2	☐
Availability of cashiers when required.	☐	☐	☐	☐	☐	3	☐
The attitude of staff when approached with your questions/queries.	☐	☐	☐	☐	☐	4	☐
The ability of staff to explain the Bank's services.	☐	☐	☐	☐	☐	5	☐
Being greeted with a smile when you visit the Branch.	☐	☐	☐	☐	☐	6	☐
Speed with which telephone is answered.	☐	☐	☐	☐	☐	7	☐
Tidiness and appearance of staff working area.	☐	☐	☐	☐	☐	8	☐
Length of time taken to get correct person on the telephone.	☐	☐	☐	☐	☐	9	☐
Speed of service in the Branch.	☐	☐	☐	☐	☐	10	☐
Prompt follow-up to written or telephone requests.	☐	☐	☐	☐	☐	11	☐
General appearance of Branch staff.	☐	☐	☐	☐	☐	12	☐
Availability of Manager or Assistant Manager when required.	☐	☐	☐	☐	☐	13	☐
Service areas clearly identified.	☐	☐	☐	☐	☐	14	☐
Ability to conduct business in private.	☐	☐	☐	☐	☐	15	☐
Effectiveness of queuing system in operation.	☐	☐	☐	☐	☐	16	☐
Availability of customer service staff when required.	☐	☐	☐	☐	☐	17	☐
Tidiness and appearance of customer area.	☐	☐	☐	☐	☐	18	☐
Staff introducing themselves by name on the telephone.	☐	☐	☐	☐	☐	19	☐

TICK FIVE BOXES ONLY

QUESTION 5 From the list of statements in question 4, please tick which 5 are most important to you?

QUESTION 6 Has the quality of service provided by this branch changed over the last 6 months? Is it now

Much worse ☐ 1 Slightly worse ☐ 2 Unchanged ☐ 3 Slightly better ☐ 4 Much better ☐ 5

QUESTION 7 Would you recommend this branch to a friend? YES ☐ 1 NO ☐ 2

QUESTION 8 Are there any other comments you would like to make?

Kindly return your completed questionnaire in the envelope provided — no stamp needed — we'll pay the postage. Thank you for your assistance.

Source: Bank of Ireland 1991

CUSTOMER COMPLAINTS ADMINISTRATION

Customer complaints checklist

Place tick in appropriate box	Yes	No
• Is there a house policy regarding customer complaints?	☐	☐
• Is there a standardised complaint response procedure?	☐	☐
• Are complaints procedure documents used for customer complaint calls?	☐	☐
• Are complaints recorded?	☐	☐
• Are complaints analysed regularly for management action?	☐	☐
• Is the cost of the complaint response procedure known?	☐	☐
• Have the problems of complaint categories over six months old been resolved?	☐	☐
• Has sub-contract liability for customer complaints been apportioned?	☐	☐
• Are staff sympathetic towards customer complaint problems?	☐	☐
• Have staff been trained to deal with angry customers?	☐	☐

If the answer to each question is Yes, OK. If No, do something about it.

Source: Katz 1987

A SAMPLE OF AN INTERNAL CLIENT SURVEY

1. Please indicate by circling your choice, the degree to which the following departments/subsidiaries are responsible for your business needs:

	To a very great extent	To a reasonably satisfactory extent	To a less than satisfactory extent	To a limited extent	Not applicable
	1	2	3	4	5
a.	______	______	______	______	______
b.	______	______	______	______	______
c.	______	______	______	______	______
d.	______	______	______	______	______

Comments

2. To what extent does the quality of each department/subsidiary of services meet your requirements?

	To a very great extent	To a reasonably satisfactory extent	To a less than satisfactory extent	To a limited extent	Not applicable
	1	2	3	4	5
a.	______	______	______	______	______
b.	______	______	______	______	______
c.	______	______	______	______	______
d.	______	______	______	______	______

Comments

3. To what extent does each department/subsidiary jointly predetermine standards for the quality of services/products they provide to you?

	To a very great extent	To a reasonably satisfactory extent	To a less than satisfactory extent	To a limited extent	Not applicable
	1	2	3	4	5
a.	______	______	______	______	______
b.	______	______	______	______	______
c.	______	______	______	______	______
d.	______	______	______	______	______

Comments

4. To what extent are service commitments delivered to you in a timely manner?

	To a very great extent	To a reasonably satisfactory extent	To a less than satisfactory extent	To a limited extent	Not applicable
	1	2	3	4	5
a.	______	______	______	______	______
b.	______	______	______	______	______
c.	______	______	______	______	______
d.	______	______	______	______	______

Comments

5. When you have a problem outside of your department/subsidiary, you know who to go to for a solution?

	Definitely agree	Inclined to agree	Inclined to disagree	Definitely disagree	Undecided; don't know
	1	2	3	4	5
a.	____	____	____	____	____
b.	____	____	____	____	____
c.	____	____	____	____	____
d.	____	____	____	____	____

Comments, if any:

6. To what extent do you understand the workers of the departments/ subsidiaries, e.g. how they affect you and your work and how you affect their work?

	To a very great extent	To a reasonably satisfactory extent	To a less than satisfactory extent	To an extremely limited extent	Not applicable
a. ____	1	2	3	4	5
b. ____	1	2	3	4	5
c. ____	1	2	3	4	5
d. ____	1	2	3	4	5

Comments, if any:

7. To what extent do you feel that other departments/subsidiaries understand you, your problems, difficulties, and obstacles?

	To a very great extent	To a reasonably satisfactory extent	To a less than satisfactory extent	To an extremely limited extent	Not applicable
a. ____	1	2	3	4	5
b. ____	1	2	3	4	5
c. ____	1	2	3	4	5
d. ____	1	2	3	4	5

Comments, if any:

8. To what extent do you feel other departments are concerned with helping you solve your problems?

	To a very great extent	To a reasonably satisfactory extent	To a less than satisfactory extent	To an extremely limited extent	Not applicable
a. ____	1	2	3	4	5
b. ____	1	2	3	4	5
c. ____	1	2	3	4	5
d. ____	1	2	3	4	5

Comments, if any:

Source: Desatnick 1988

CUSTOMER COMPLAINT

Date received:	Received by:
How received: Caller ☐ Telephone ☐ Letter (attach copy) ☐	
Name of complainant	
Company name	
Company address	Telephone number (inc. STD code and ext.)
Goods complained of (if any)	Value
Proof of purchase	Were goods produced/returned?
Date: Document: Number:	
Nature of complaint	
Category of complaint: A ☐ B ☐ C ☐ D ☐	
Immediate action taken	
Further action required	

Source: Peel 1987

IMPROVING PERSONAL PERFORMANCE

Question: What is the best format for do-it-yourself evaluation of performance in answering the phone?

Day-to-day activities are rarely appraised by the persons involved. Their performance is an integral part of routine and habit. It rests with management to initiate an inward look at periodic intervals. A cycle of three months makes sure that performance standards are not allowed to deteriorate. The following checklist helps service personnel to monitor their own performance:

The incoming call checklist

Please tick in appropriate box	Yes	No
Yesterday:		
• Did I smile when picking up the receiver?	☐	☐
• Did I start the conversation by giving the name of the department, and my name?	☐	☐
• Did I establish the caller's requirements correctly?	☐	☐
• Did I give all the required information correctly?	☐	☐
• When applicable, did I arrange to get necessary information and then call back to give it to the caller?	☐	☐
• Did I warn callers first, when I had put the phone down for a moment in the middle of a call?	☐	☐
• Did I deal with caller's objections and complaints in an effective way?	☐	☐
• Did I close the call properly?	☐	☐
• Did I leave my name with every caller requiring action?	☐	☐

RATE YOUR CUSTOMER SERVICE STANDARDS

Mark in each your score from one to five.

	1	2	3	4	5	
Chief Executive delegates Customer Service						Chief Executive committed to Customer Service
Senior Management concentrates on budgets/figures						Senior Management talk to staff, customers, suppliers
Staff attitudes stop progress in Customer Care						Management force Customer Care improvement
Every department driving for profit or results						Every department driving for Customer Care
Only Sales and Marketing have customers						Every Department has customers
Customer Service is the Customer Service Department's problem						Customer Service is everyone's responsibility
It's alright to upset a few customers						Every customer should be treated right first time
The Manager's role is instruction and discipline						The Manager's role is to coach and support a team
Systems are run for the company						Systems are run for the Customer
People are trained to follow a formula						People are trained to respond situationally
No one ever mentions Customer Service						Customer Service is a continuous communication and training topic
Total	➩	➩	➩	➩	➩	➩ ☐ Grand Total

A SAMPLE EMPLOYEE OPINION SURVEY

Please record your attitude toward each statement by drawing a circle around the answer you choose. For example, please draw a circle around:

- *Definitely Agree*: if the statement definitely expresses how you feel about the matter.
- *Inclined to Agree*: if you are not definite, but think that this statement tends to express how you feel about the matter.
- *Inclined to Disagree*: if you are not definite, but think that the statement does not tend to express how you feel about the matter.
- *Definitely Disagree*: if the statement definitely does *not* express how you feel about the matter.

The statements in this part of the questionnaire express a wide range of feelings that a person might have about his job. In recording *your* feelings, please indicate what *you* think about *your* job — what you like about it, and what you dislike. You can do this by showing how much you agree or disagree with each statement.

In some of the following statements, the term 'department' is used. This refers to the office or branch where you work. Your manager is the person to whom you report.

1. I enjoy my work here.
   Definitely Agree — Inclined to Agree — Inclined to Disagree — Definitely Disagree
2. I am satisfied with the salary I now receive.
   Definitely Agree — Inclined to Agree — Inclined to Disagree — Definitely Disagree
3. Decorations and furnishings in my office are not in keeping with the kind of job I have.
   Definitely Agree — Inclined to Agree — Inclined to Disagree — Definitely Disagree
4. I am satisfied with my chances to be promoted to a better position (higher level) in the future.
   Definitely Agree — Inclined to Agree — Inclined to Disagree — Definitely Disagree
5. Too frequently I am kept in the dark about what goes on around here.
   Definitely Agree — Inclined to Agree — Inclined to Disagree — Definitely Disagree
6. The people I work with always have the time to give the information I need to do my work.
   Definitely Agree — Inclined to Agree — Inclined to Disagree — Definitely Disagree
7. I am satisfied with the extent to which the work I am now doing is receiving the recognition and respect of my associates.
   Definitely Agree — Inclined to Agree — Inclined to Disagree — Definitely Disagree
8. I get a great deal of satisfaction out of my work because it means being connected with a successful (profitable, effective) operation.
   Definitely Agree — Inclined to Agree — Inclined to Disagree — Definitely Disagree
9. I feel that my job is not classified at a sufficiently high position level.
   Definitely Agree — Inclined to Agree — Inclined to Disagree — Definitely Disagree
10. We need more privacy in our office arrangement.
    Definitely Agree — Inclined to Agree — Inclined to Disagree — Definitely Disagree
11. I am satisfied with my chances in the future to do work which offers good opportunities for continued growth in my profession or technical specialty.
    Definitely Agree — Inclined to Agree — Inclined to Disagree — Definitely Disagree
12. My manager always lets me know beforehand of changes that affect my work.
    Definitely Agree — Inclined to Agree — Inclined to Disagree — Definitely Disagree
13. The people I work with often are too busy to help me when I need the service they are supposed to provide.
    Definitely Agree — Inclined to Agree — Inclined to Disagree — Definitely Disagree
14. I am satisfied with the extent to which the work I am now doing will probably have a significant influence on the future of the Company (or of my department).
    Definitely Agree — Inclined to Agree — Inclined to Disagree — Definitely Disagree
15. I get a great deal of satisfaction out of my work because of the good products we make (or good services we render) in this department.
    Definitely Agree — Inclined to Agree — Inclined to Disagree — Definitely Disagree

Source: Desatnick 1988

AN EXIT INTERVIEW QUESTIONNAIRE

To be completed for each terminated employee.
Name ____________________
Office ____________________
Job title ____________________
Date interviewed ____________________
By ____________________
Termination date ____________________
Supervisor ____________________

We want to make this organization a better place in which to work, and we need your help to do so. Would you please spare five or ten minutes of your time to answer a few questions?

1. If you accepted another job, what does that job offer that your job here did not?
2. What were the factors which contributed to your accepting a job with this company? Were these expectations realized? If not, why not? Has that changed in your present job?
3. Was the job you held accurately described when you were hired? To what extent do you feel that your skills were utilized?
4. What constructive comments would you have for management with regard to making our organization a better place in which to work?
5. What are some of the factors which helped to make your employment enjoyable for those parts that you liked?
6. Would you recommend this company to a friend as a place to work? If yes, why? If no, why not?
7. Was your decision to leave influenced by any of the following? Please check all those that are applicable.

Leaving the city	______
Returning to school	______
Health reasons	______
Family circumstances	______
Retirement	______
Secured a better position	______
Dissatisfied with:	
type of work	______
working conditions	______
salary	______
supervision	______
other (please specify)	____________________

8. What did you think of the following in your job or your department? (please pick one)

	Excellent	Good	Fair	Poor
• Orientation to job	____	____	____	____
• Physical working conditions	____	____	____	____
• Equipment provided	____	____	____	____
• Adequacy of training	____	____	____	____
• Fellow workers	____	____	____	____
• Cooperation within the department	____	____	____	____
• Cooperation with other departments	____	____	____	____
• Workload	____	____	____	____

9. What was your attitude regarding your supervisor/manager?

	Excellent	Good	Fair	Poor
• Demonstrates fair and equal treatment	____	____	____	____
• Provides recognition on the job	____	____	____	____
• Resolves complaints and problems	____	____	____	____
• Follows consistent policies and practices	____	____	____	____
• Informs employees on matters that directly relate to their jobs	____	____	____	____
• Encourages feedback and welcomes suggestions	____	____	____	____
• Knowledgeable regarding performance and accomplishment of his employees	____	____	____	____
• Expresses instructions clearly	____	____	____	____
• Develops cooperation	____	____	____	____

10. What is your opinion regarding the following? (choose one)

	Excellent	Good	Fair	Poor
• Your salary	____	____	____	____
• Opportunity for advancement	____	____	____	____
• Performance appraisal	____	____	____	____
• Company polices	____	____	____	____

(If fair or poor, tell why.)

Source: Desatnick 1988

EMPLOYEE ASSESSMENT SURVEY

We continually look for ways to improve service to our customers. Your ideas and suggestions will be a very important part of our plans to provide better service. Please complete this questionnaire and be candid about your observations. Because you have regular contact with our customers, you know best where *improvement* is needed.

Please return the questionnaire, unsigned, to (*name/department*) by (*date*). Thank you for your suggestions.

1. How do you feel our service compares with the service provided by (primary competitors)?

2. What problems do you face in delivering high-quality service to your customers?

3. Which one of the above is the greatest problem you face?

4. What feedback regarding service do you receive from your customers most frequently?

5. What changes would you make to improve service to your customers?

6. What gets in the way of you delivering good service?

7. Why do you think customers leave?

Thank you for your help.

Chief Service Officer

Source: Liswood 1990

SAMPLE 'QUALITY SERVICE' QUESTIONNAIRE FOR EMPLOYEES

Sample Employee Questionnaire: Developing a Definition of Quality

In our continuing efforts to improve service quality, we will attempt to define what 'quality of service' means to (institution name). This definition will become the foundation for our service quality program.

Your opinions are important! Please complete this brief questionnaire and return it, unsigned, to (name/department) by (date).

1. Which of the following are important elements in your definition of quality service?

	Important Part of Quality Definition	Not an Important Part of Quality Definition
Promptness of service	☐	☐
Professionalism of staff	☐	☐
Accuracy of information	☐	☐
Clarity of statements/billings	☐	☐
Courtesy of staff	☐	☐
Knowledge of staff	☐	☐
Reputation of company	☐	☐
Friendliness of staff	☐	☐
Availability of brochures/information	☐	☐
Hours of operation	☐	☐
Other ____________	☐	☐

2. If you were responsible for developing a written definition of quality for (name), what definition would you recommend?

____

____

Source: Liswood 1990

STAFF EDUCATION CHECKLIST

Place tick in appropriate box	Yes	No
• Is there specific staff training in telephone techniques?	☐	☐
• Is there an induction training programme for new employees?	☐	☐
• Are staff trained for face-to-face meetings with customers and the public?	☐	☐
• Do customer needs take priority over in-house company activities?	☐	☐
• Is management action taken when employee irritation with customers is identified?	☐	☐
• Is there specific staff training in respect of PR – adverse and favourable – that is generated by employee conduct?	☐	☐
• Do the staff like the customers?	☐	☐
• Do the staff know about customer levels of expectation?	☐	☐
• Are there specific incentives to motivate staff?	☐	☐
• Do staff have job satisfaction?	☐	☐

If the answer to each question is Yes, OK. If No, do something about it.

Source: Katz 1987

Bibliography

The items starred (*) are particularly recommended.

Administration Yearbook and Diary (1989) Institute of Public Administration Dublin

Albrecht, K. (1988) *At America's Service* Homewood, Illinois: Dow Jones-Irwin

Albrecht, K. and Bradford, L.J. (1990) *The Service Advantage: How to identify and fulfill customer needs* Homewood, Illinois: Dow Jones-Irwin

Albrecht, L. and Zemke, R. (1985) *Service America! Doing Business in the New Economy* Homewood, Illinois: Dow Jones-Irwin

Atkinson, P.E. (1990) *Creating Culture Change: The key to successful total quality management* Bedford, England: IFS publications

Aubrey, C.A. (1988) *Quality Management in Financial Services* Wheaton: Hitchcock Publishing Company

Baker, K.R. (1974) *Introduction to Sequencing and Scheduling* New York: Wiley

Berry, L.L. and Bennett, D.R. and Brown, C.W. (1989) *Service Quality: A Profit Strategy For Financial Institutions* Homewood, Illinois: Dow Jones-Irwin

Blumberg, D.F. (1991) *Managing Service as a Strategic Profit Center* New York: McGraw-Hill

Bone, D. and Griggs, R. (1989) *Quality At Work* London: Kogan Page

Brown, A. (1989) *Customer Care Management* Oxford: Heinemann Professional Publishing

Butterfield, R.W. (1991) *Quality Service Pure and Simple* Wisconsin: ASQC Quality Press

Buzzell, R. and Gale, B.T. (1987) *The PIMS Principles — Linking Strategy to Performance* New York: The Free Press 1987

CERT (1990) *A Practical Guide to Customer Relations* Dublin: Mount Salus Press

Caplan, F. (1990) *The Quality System* (2nd ed.) Pennsylvania: Chilton Book Company

Cava, R. (1990) *Dealing with Difficult People* London: Judy Piatkus

Cooper, R. and Kaplan, R.S. (1991) *The Design of Cost Management Systems* New Jersey: Prentice Hall

Crosby, P.B. (1979) *Quality is Free* New York: McGraw-Hill Book Company

Crosby, P.B. (1986) *Running Things, The Art of Making Things Happen* Singapore: McGraw-Hill

Crosby, P.B. (1987) *Quality Without Tears* (3rd ed.) Singapore: McGraw-Hill

D'Egidio, F. (1990) *The Service Era Leadership in a Global Environment* Cambridge, Massachusetts: Productivity Press

Davidow, W.H. and Uttal, B. (1989) *Total Customer Service The Ultimate Weapon* New York: Harper & Row

Deming (1986) *Out of the Crisis* Cambridge, Massachusetts Institute of Technology

Deming (1982) *Quality, Productivity and Competitive Position* Cambridge, Massachusetts Institute of Technology

Denton, D.K. (1989) *Quality Service* Houston, Texas: Gulf Publishing Company

Desatnick, R.L. (1988) *Managing to Keep the Customer* London: Jossey-Bass

DiPrimio, A. (1987) *Quality Assurance in Service Organisations* Pennsylvania: Chilton Book Company

*Dixon, N.F. (1976) *On the Psychology of Military Incompetence* London: Cape

Eldridge, L. (1983) *Bank America*

European Organization for Quality (1988) *Methods for Determining Suitable AQL Values* Bern, Switzerland EOQ

European Organization for Quality (1990) *Proceedings of 34th Annual Conference. 'Winning Through Quality'* Dublin: Irish Quality Association

European Organization for Quality (1991) *Proceedings of 35th Annual Conference. 'The Human Factor in Quality Management'* Prague: EOQ

Feigenbaum, A.V. (1983) *Total Quality Control* (3rd ed.) New York: McGraw-Hill

*Gitlow, H.S. and S.J. (1987) *The Deming Guide to Quality and Competitive Position* Englewood Cliffs, NJ: Prentice Hall

Goldratt, E.M. (1990) *The Haystack Syndrome: Sifting Information out of the Data Ocean* New York: North River Press

Goonroos, C. (1988) *Service Quality: The Six Criteria of Good Perceived Service Quality* Review of Business St John's University

Gummesson, E. (1989) 'Nine Lessons on Service Quality' in *Total Quality Management* vol 1, no 2

Hall, S.J. (1990) *Quality Assurance in the Hospitality Industry* Wisconsin: ASQC Quality Press

Harrington, H.J. (1987) *The Improvement Process, How America's Leading Companies Improve Quality* New York: McGraw-Hill

Hellriegel, Slocum, Woodman (1989) *Organizational Behaviour* St Paul, Minnesota: West

Heskett, J.L. and Sasser, W.E. and Hart, C.W. (1990) *Service Breakthroughs, Changing the Rules of the Game* New York: The Free Press

Hickman, C.R. and Silva, M.A. (1986) *Creating Excellence* London: Unwin
Hollins, G. and Hollins, B. (1991) *Total Design Managing the design process in the service sector* London: Pitman
Ishikawa, K. (1976) *Guide to Quality Control* Tokyo: Asia Productivity Organisation
*ISO (1991) *Quality Management and Quality System elements Part 2. Guidelines for services (ISO 9004–2)* Geneva ISO
Juran, J.M. and Gryna, F.M. and Bingham, R.S. (eds) (1979) *Quality Control Handbook* (3rd ed.) New York: McGraw-Hill
Juran, J.M. and Gryna, F.M. (1980) *Quality Planning and Analysis* New York: McGraw-Hill
Katz, B. (1987) *How to turn Customer Service into Customer Sales* Aldershot, England: Gower
Lammermeyr, H.U. (1990) *Human Relations the Key to Quality* Wisconsin: ASQC Quality Press
Lapin, L. (1988) *Quantitative methods for Business Decisions* Orlando, Florido: Harcourt Brace Jovanovich
Lash, L.M. (1989) *The Complete Guide to Customer Service* New York: Wiley
Latzko, W.J. (1986) *Quality and Productivity for Bankers and Financial Managers* New York: Marcel Dekker
Lefevre, H.L. (1989) *Quality Service Pays Six Keys to Success!* Wisconsin: ASQC Quality Press
Lipsey (1963) *An Introduction to Positive Economics* London Weidenfeld and Nicholson
Liswood, L.A. (1990) *Serving Them Right* New York: Harper Business
Lock, D. and Smith, D.J. (eds) (1990) *Gower Handbook of Quality Management* Aldershot, England: Gower
Love, J.F. (1988) *McDonald's: Behind the Arches* London: Bantam
Martin, W.B. (1989) *Managing Quality Customer Service* London: Kogan Page
Masing, W. (1990) *Some Considerations on Quality Costs. 34th EOQ Conference Proceedings* Dublin: Irish Quality Association
NSAI Annual Conference Proceedings May 1991 Dublin: NSAI
McPherson, James (1990) *Battlecry of Freedom* Harmondsworth: Penguin
Møller, C. (1988) *Personal Quality* TMI International
Murphy, J.A. (1988) *Quality in Practice* Dublin: Gill and Macmillan
*Oakland, J.S. (1989) *Total Quality Management* Oxford: Heinemann Educational Books
J.W. O'Hagan *The Economy of Ireland* (6th ed.) 1991 Dublin: Irish Management Institute
Peel, M. (1987) *Customer Service, How to Achieve Total Customer Satisfaction* London: Kogan Page
Peters, T. (1986) *Quality!* Palo Alto, California: Peters Organisation
*Peters, T. (1987) *Thriving on Chaos* London: Macmillan
Peters, T.J. and Waterman, R.H. (1982) *In Search of Excellence: Lessons from America's Best Run Companies* New York: Harper & Row

Price F. (1984) *Right First Time* Aldershot, England: Gower
Quinn, F. (1990) *Crowning the Customer* Dublin: The O'Brien Press
Quinn, M. and Humble, J. (1991) *Service: The New Competitive Edge, a Survey of executive opinion of senior managers in Ireland* Dublin Digital Equipment Corporation
*Reichheld, F.F. and Sasser Jr, W.E. (Sept/Oct 1990, Jan/Feb 1991) *Zero Defections: Quality Comes to Services* Harvard Business Review
Rosander, A.C. (1985) *Applications of Quality Control in the Service Industries* Wisconsin: ASQC Quality Press
Rosander, A.C. (1989) *The Quest For Quality in Services* Wisconsin: ASQC Quality Press
Sayle, A.J. (1988) *Management Audits: The Assessment of Quality Management Systems* (2nd ed.) London: Allan J. Sayle
Schein, E.H. (1989) *Organisational Culture and Leadership* Oxford: Josey-Bass
Schonberger, R.J. (1990) *Building A Chain of Customers* New York: The Free Press
Sidney, E. (1988) *Managing Recruitment* Aldershot, England: Gower
Smith, A. (1776) *An Inquiry into the Nature and Causes of the Wealth of Nations* London
Stahl, I. (1990) *Introduction to Simulation with GPSS* New Jersey: Prentice Hall
Starr, M.K. (1989) *Managing Production and Operations* New Jersey: Prentice Hall
Terkel, S. (1974) *Working* New York: Ballantine
Townsend, P.L. and Gebhardt, J.E. (1986) *Commit to Quality* New York: John Wiley and Sons
Trachtenberg, J. (1968) *The Speed System of basic mathematics* London: Pan
Wilkie, W.L. (1986) *Consumer Behavior* New York: Wiley
*Zeithaml, V.A., Parasuraman, A. and Berry, L.L. (1990) *Delivering Quality Service, Balancing Customer Perceptions and Expectations* New York: The Free Press

Index